New Zealand's Economic Native Plants

New Zealand's
ECONOMIC
NATIVE PLANTS

R.C. Cooper and R.C. Cambie

Auckland
OXFORD UNIVERSITY PRESS
Melbourne Oxford New York

Oxford University Press
Oxford University Press, Walton Street, Oxford OX2 6DP

OXFORD   NEW YORK   TORONTO
DELHI   BOMBAY   CALCUTTA   MADRAS   KARACHI
PETALING JAYA   SINGAPORE   HONG KONG   TOKYO
NAIROBI   DAR ES SALAAM   CAPE TOWN
MELBOURNE   AUCKLAND
and associated companies in
BERLIN   IBADAN

OXFORD is a trade mark of Oxford University Press

First published 1991
© R. C. Cooper and R. C. Cambie 1991

ISBN 0 19 558229 2

Cover illustration by Chris O'Brien
Photoset in Times by Rennies Illustrations Ltd
and printed in Hong Kong
Published by Oxford University Press
1A Matai Road, Greenlane, Auckland 5, New Zealand

Contents

Plates

Preface

I am delighted to see that this book, *New Zealand's Economic Native Plants*, is being revised at a time of resurgent interest in the uses of native plants. The book represents an important source for people interested in the story behind our native plants — from single cell organisms to giant podocarps.

For the entrepreneur, the scientist, and the gardener, it will be a store of information to be consulted frequently. The references are comprehensive, and frequent mention of the connections between plants and individuals and events in the history of New Zealand lends a wider context and provides a deeper understanding of the complexity and richness of the ecosystem.

New Zealand's Economic Native Plants will fill a niche in the libraries of New Zealanders with even a cursory interest in the plants in their gardens and in the bush. More importantly, it will provide an easy-to-read guide to those who are just beginning to be aware of the variety, complexity, and value of our native plants.

Simon Upton
Minister of Research, Science, and the Environment
Wellington, August 1991

Acknowledgements

We are very grateful to the botanists, chemists and librarians who assisted us while we were preparing this book, and in particular we thank:

Dr J. E. Braggins, University of Auckland, Private Bag, Auckand 1, New Zealand.

Dr Peter Buchanan, DSIR, Mt Albert Division, Private Bag, Auckland, New Zealand.

Professor T. Carrick Chambers, Director, Royal Botanic Gardens, Sydney, NSW 2000, Australia.

Dr Brian T. Coffey and Associates, 1 Vesty Place, Hillcrest, Hamilton, New Zealand.

Dr Vivienne Cassie Cooper, DSIR Land Resources, Ruakura Agricultural Research Centre, Hamilton, New Zealand.

Ms Lisa Crozier, Forest Research Institute, Box 31-011, Christchurch, New Zealand.

Dr Janet M. Davidson, National Museum, Wellington, New Zealand.

Dr Lionel Gilbert, 151 Brown Street, Armidale, NSW 2350, Australia.

Dr Margaret Gordon, Victoria University of Wellington, PO Box 600, Wellington, New Zealand.

Miss Colleen Hamill, Department of Chemistry, University of Auckland.

Dr Warwick Harris, DSIR Land Resources, Botany Institute, Private Bag, Christchurch, New Zealand.

Dr John H. Harvey, 32 Christchurch Street East, Frome, Somerset BA11 1QH, England.

Dr Eric McKenzie, DSIR, Mt Albert Division, Private Bag, Auckland, New Zealand.

Mr Lawrie Metcalf, 117 Moana Street, Rosedale, Invercargill, New Zealand.

Dr Neville Moar, 8 Maurice Street, Lincoln, New Zealand.

Dr Peter C. Molan, Department of Biological Sciences, University of Waikato, Private Bag, Hamilton, New Zealand.

Mrs Sheila Natusch, 46 Ohiro Bay Parade, Wellington, New Zealand.

Dr E. Charles Nelson, 14 Connaught Parade, Phibsborough, Dublin 7, Republic of Ireland.

Dr Wendy Nelson, National Museum, Wellington, New Zealand.

Mrs Lynley Nesbitt, Library, Botany Division, DSIR, Lincoln, New Zealand.

Professor Norton H. Nickerson, Tufts University, Medford, MA 02155, USA.

Dr Murray Parsons, DSIR Land Resources, Botany Institute, Private Bag, Christchurch, New Zealand.

Dr Tim Pickering, Mahanga Bay Shellfish Hatchery, MAFish, Fisheries Research Centre, Box 297, Wellington, New Zealand.
Dr A.D. Thomson, Centre for Studies of NZ Science History, 5 Karitane Drive, Christchurch 2, New Zealand.
Prof. Dr G. Wagenitz, Systematisch-Geobotanisches Institut der Universität, Untere Karspule 2, 3400 Göttingen, Germany.
Dr Rod Wallace, Anthropology Department, University of Auckland, Private Bag, Auckland 1, New Zealand.
Mr Anthony Wright, Botanist, Auckland Institute and Museum, Private Bag, Auckland 1, New Zealand.

We also wish to thank the Auckland Institute and Museum for permission to use photos, drawings, and paintings.

1

Introduction

1.1 The New Zealand Vascular Flora

In the first edition of this book the number of species of native vascular plants was estimated at 2000, comprising:

Ferns and fern allies		166
Seed plants:	Gymnosperms	20
	Monocotyledons, excluding grasses	339
	Grasses	171
	Dicotyledons	1304
		Total 2000

The latest estimate is 2200 species, of which possibly 300 are known but undescribed.[1,2] As well as vascular plants, the native flora includes seaweeds, marine and freshwater microalgae, mosses, lichens, liverworts, fungi, and so on, which are discussed later.

1.2 Features of the Flora

Leonard Cockayne described the main features of the native flora in a series of articles in the *Lyttelton Times* during April and May 1906. The articles also appeared in the *Dunedin Evening Star* and *Auckland Star* and, in 1910, were reproduced in book form as *New Zealand Plants and their Story*. Revised editions appeared in 1919, 1927, and 1967, the last by Eric Godley. These introduced generations of New Zealanders to their native plants.

The latest book on the distinctive features of the indigenous flora, *Forest Vines to Snow Tussocks: The Story of New Zealand Plants* by John Dawson,[1] was published by Victoria University Press in 1988 and was based on Cockayne's now outdated classic.

From this wealth of information about endemism, hybridism, divaricating habit, juvenile life form, filmy ferns, epiphytes, vegetable sheep, and so on, it seems that one of the main features of the flora is its age and origin. The late Sir Charles Fleming, doyen of New Zealand geologists and naturalists, concluded a talk on the history of life in New Zealand forests with these words:

> The oldest life in New Zealand forests goes back more than 100 million years to Gondwanaland. Other southern lands started with a similar heritage, but have differentiated owing to their reduced isolation from the Northern Hemisphere (South America and Africa), or the Late Tertiary desiccation that triggered off the evolution of savannahs, of eucalypts in Australia and the veldt of Africa, both of which countries retained more Proteaceae than New Zealand. As a result New Zealand forests are most like the Mesozoic forests of Gondwanaland, and are so regarded by Northern Hemisphere biologists who visit them in pilgrimage. We are the custodians of a New Zealand heritage of world interest.[3]

1.3 Use of the Flora in pre-European Times

The Māori made use of many native plants; the Tuhoe people, for example, living in the Urewera where conditions were not favourable for kūmara cultivation, were compelled to rely upon minor products of the bush. Elsdon Best (1903) compiled a list of the native plant products consumed by the Tuhoe as food, and it is clear that they had recourse to most plants in the Urewera flora.[4]

Later in this book we give brief lists of the plants employed by the Māori in gardening (Chapter 4), as timber (Chapter 5), for plaiting and weaving (Chapter 6), and so on. However, it must be emphasized that other authors have provided much fuller, authoritative accounts; our notes on past Māori use are intended merely to introduce discussion of a plant or group of plants with a view to its possible future utilization.

At this point it must be observed that there were marked differences in the economies of tribes; for example Jeanne Goulding (1971) found leaves of Māori onion (*Bulbinella*) and neinei (*Dracophyllum*) in woven fragments from archaeological excavations in Fiordland, while elsewhere native flax (*Phormium*), pīngao (*Desmoschoenus*), and similar plants were used for weaving.[5]

According to Janet Davidson (1983) social networks developed early on,[6,7] and there were frequent references to them. Edward Shortland (1851) described trading or barter in such items as the fragrant oil of the taramea (*Aciphylla*) from the south, and tōtara (*Podocarpus*) from the north. Tōtara logs were used for making canoes, and tōtara bark was used to make the casks for storing mutton birds.[8]

H. Beattie (1920), an authority on the South Island Māori people, described how they exchanged mutton birds from the south for such articles as taro

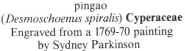

pīngao
(*Desmoschoenus spiralis*) **Cyperaceae**
Engraved from a 1769-70 painting
by Sydney Parkinson

taramea (speargrass)
(*Aciphylla squarrosa*) **Apiaceae**
Engraved from a 1769-70 painting
by Sydney Parkinson

(*Colocasia*) and gourds or hue (*Lagenaria*) from the north.[9] Peter Buck (1923) mentioned inland tribes obtaining prepared material of pīngao from their coastal friends.[10]

B.F. and H. Leach (1979) concluded from archaeological investigations that Polynesian settlers, who lived in eastern Palliser Bay between about A.D. 1100 and 1400, obtained rocks for tools and weapons from as many as fifteen different places as far afield as Northland, Central Otago, and the West Coast of the South Island. Presumably they exchanged kūmara and gourds for the stones, but definite evidence for barter was not found in the excavations.[11] It must be added that barter in the commercial sense is not the best term. Polynesian and Maori custom seems to be to exchange goods such as food, cloaks and mats as gifts when groups meet.

1.4 Use of Native Plants Today

All plants, native and exotic, have value. All remove CO_2 from the atmosphere and release oxygen. Some play a part in nitrogen production, others in nitrogen fixation. In salt marshes and mangrove swamps, they play an essential role in providing food for shellfish and fish, and in cleansing our coastal waters. In the hills, they play another vital role in slowing runoff, holding water, and checking erosion. Throughout the country, they have scenic value, and are an essential ingredient of our tourist industry. None are useless. All need our protection.

We must begin this book with a caveat about further damage to New Zealand's unique flora. The reduction of the native flora, by fire, farming and forestry, has reached a stage that is unacceptable to any reasonable person concerned with retaining the national heritage. Any attempt to utilize native plants for commercial purposes must have regard to the fact that New Zealand flora is now in the endangered category.

1. Dawson, J. *Forest Vines to Snow Tussocks: The Story of New Zealand Plants* Wellington (Victoria University Press) 1988 264p.
2. *Report of the Review Committee on Biosystematics and Ecological Science.* DSIR Science Activity Area Reviews 4 (1989) 49p.
3. Fleming, C.A. (1977) 22 *N.Z.J. Forestry* 249, reprinted in (1978) 210 Forest and Bird 2.
4. Best, E. (1903) 35 *Trans. N.Z. Inst* 45.
5. Goulding, J.H. (1971) 8 *Rec. Auckland Inst. Mus.* 57.
6. Davidson, J.M. (1983) 91 *J. Polynes. Soc.* 291.
7. Davidson, J.M. *The Prehistory of New Zealand* Auckland (Longman Paul) 1984 270p.
8. Shortland, E. *The Southern Districts of New Zealand.* London (Longman, Brown etc.) 1851; reprinted by Capper Press 1974 316p.
9. Beattie, H. (1920) 52 *Trans. N.Z. Inst.* 53.
10. Buck, P.H. (1923) 54 *Trans. N.Z. Inst.* 705.
11. Leach, B.F., Leach, H. (eds) *Prehistoric Man in Palliser Bay* Wellington (National Museum of New Zealand Bull. 21) 1979 vi, 272p.

2

History of the Utilization of the Flora

2.1 First European Visitors

In 1769-70, during Captain James Cook's first visit to New Zealand, Mr Joseph Banks listed about fifteen useful plants in his journal.

> . . . as People who had been long at Sea [we] found great benefit in the article of health by eating plentifully of Wild Celery, & a kind of Cresses which grew everywhere abundantly near the Sea Side; we also once or twice met with an herb like that which the Countrey People in England call Lambs Quarters or Fat hen which we boiled instead of Greens, & once only a Cabbage tree, the cabbage of which made us one delicious meal. . .

Elsewhere Banks mentioned a 'Palm cabbage'.

Banks also noted sow thistle, garden nightshade, fern roots and one other vegetable (Pandanus) 'totally unknown in Europe, which tho eat by the Natives, no European will probably ever rellish . . .'.

He continued:

> Nor does their cultivated Grounds produce many Species of Esculent Plants, three only I have seen, Yamms, Sweet Potatoes, & Cocos . . . they cultivate often pieces of many Acres, & I believe any Ship that was to be to the Northward in the Autumn, about the time of digging them up might purchase any quantity.
>
> . . . besides these they cultivate Gourds, the fruits of which serve them to make Bottles, Jugs, etc & a very small quantity of the Chinese paper mulberry tree, the same as the inhabitants of the S. Sea Islands use to make their Garments of, this they very much value but it is so scarce with them probably having been brought from a hotter Climate & not thriving here . . .
>
> Fruits they have none, except I should reckon a few kind of insipid berries which had neither sweetness nor flavour to recommend them, & which none but the Boys took the pains to gather; the Woods however abound with excellent Timber trees fit for any kind of building . . .

From Banks's notes it is possible to recognize pōhutukawa and kahikatea trees.

> But of all the Plants we have seen among these People, that which is the most excellent in its kind, & really excels most if not all that are put to the same uses in other Countries, is the plant which serves them instead of hemp & flax . . .

Banks had no name at this time for New Zealand flax (*Phormium tenax*); he described the split leaves used to make nets and the snow-white fibres used to make fine clothes, and concluded that the plant would be a great acquisition to England where it might thrive with little trouble.[1]

Publication of Banks's journal was delayed until Sir Joseph Hooker edited it in 1896, but Banks made the manuscripts, specimens, and drawings from the voyage available to others during his lifetime, at his house in Soho Square, London.[2]

Captain Jean de Surville and members of his crew recorded some of the useful plants of New Zealand during the visit of the *St. Jean Baptiste* in 1769-70, but again, publication of the journals was delayed for many years. They noticed 'young potato crops', presumably sweet potatoes, stores of fern root, and cultivations of karaka trees. They described two kinds of watercress and a wild parsley or celery, which greatly benefited the sailors who were suffering from scurvy, and they mentioned the purchase of some lamp oil, apparently made from tītoki berries.[3,4]

In Captain Cook's account of the second voyage, published in 1777, he wrote that future explorers would find that the most useful New Zealand plants were flax (*Phormium tenax*) for cordage, rimu (*Dacrydium cupressinum*) for spruce beer, and mānuka (*Leptospermum scoparium*) for tea. He included drawings of the three plants.[5]

George Forster, who had accompanied his father, J.R. Forster, on the second voyage, published a two-volume work, *A Voyage around the World in His Britannic Majesty's Sloop, Resolution*, shortly before the issue of Cook's official account. In it, George Forster wrote of mānuka tea, the fine aromatic flavour of which became very bitter on the second filling of the teapot. Of beer brewed with rimu twigs, he recorded that too great an astringency was corrected by an equal amount of tea-tree, the only fault of the beer being the nausea it caused on an empty stomach. He described the discovery of the tōī or mountain cabbage tree on the heights above Dusky Sound, remarking that the central shoot, when quite tender, tasted something like an almond kernel with a little of the flavour of cabbage.[6]

In 1786, George Forster published his doctoral thesis on esculent (food) plants of the Southern Ocean. In Hatch's 1986 English translations of Forster's Latin, we learn that the sailors had *Lepidium oleraceum* as a vegetable, with

Apium and *Tetragonia*, every day while in Queen Charlotte Sound. 'The taste (of *Lepidium*) is pleasant though somewhat bitter, similar to that of spinach or lettuce. A mild but rather flatulent laxative'. *Lepidium* is Cook's scurvy-grass, *Apium* is native celery, and *Tetragonia* is New Zealand spinach.

Of *Sonchus oleraceus*, Forster's comment was: 'its delicate stems and young leaves made a salad for our enjoyment'. Of *Cyathea medullaris* he wrote that the Māori 'roast and eat the pith of the root and lower stem. A soft and pulpy substance, with an abundant reddish glutinous sap, it has a taste like that of turnip but rather better, and in this it approaches the pith of the sago tree.' *Sonchus* is sow thistle; *Cyathea medullaris* is black tree-fern or mamaku.

Of *Coriaria sarmentosa*, Forster wrote 'the natives of this region, particularly those who lived near Dusky Bay, collected and ate the berry-like fruits of this shrub'; he did not warn readers that the seeds are very poisonous. Forster's Latin text on *Solanum aviculare* may be translated: 'The acid yellow almost nauseous fruits are smaller than a plum and are eaten greedily by the New Zealanders and by small birds, and were not entirely rejected by our people.' The species of *Coriaria* are known as tutu, and *Solanum aviculare* is poroporo. *Avicennia resinifera* was included as an esculent plant by mistake. It was believed at that time that kauri gum, which was chewed by the Māori, came from the mangrove.[7,8]

For the third voyage, Captain Cook had William Anderson as surgeon and naturalist. Anderson had served on the second voyage and, after experiencing the vagaries of scientists, Cook probably preferred a shipmate. The expedition visited Queen Charlotte Sound from 12 to 26 February 1777, and Anderson collected plants. Unfortunately, he died at sea on 3 June 1778, and his plants were overlooked, severely damaged by insects, and subsequently dismissed by J.D. Hooker as small and unimportant. David Nelson, Anderson's assistant, is credited with the collection of seeds that reached England in 1780. Anderson's journal was used extensively in the official account of the third voyage.[9,10,11,12]

The first sealing gang from Sydney was established at Dusky Sound in November 1792 and during their stay of ten months the men built a vessel of New Zealand timbers that was later completed and launched as the *Providence*. By 1825, the New Zealand seal population had declined and the sealers obtained flax to fill their vessels.[13]

2.2 Missionaries and Settlers

Early visitors, missionaries and settlers added steadily to our knowledge of native plants; for example, Reverend William Yate, who joined the mission at the Bay of Islands in January 1828, published an account of New Zealand in 1835, in which he described common trees of the northern part of the country:

kauri (*Agathis australis*) **Araucariaceae**
From A.B. Lambert *Pinetum or Genus Pinus* (1832)

kauri, tānekaha, tōtara, rimu, kahikatea, maire, taraire, tawa, pūriri, rewarewa, kawaka, miro, tōwai, pōhutukawa, ake, kahikātoa, kohekohe, māhoe, hīnau, matai and rātā. He wrote that naval spars could be obtained, but only with much labour and expense, and recommended that those seeking cargoes of spars try the Hokianga and Whangaroa Harbours in preference to the Firth of Thames (where Cook had seen forests of kahikatea). Although there are some errors, his book must have been of great assistance to timber merchants and settlers.[14]

In 1842 Rev. W. Wade, who served as a missionary in New Zealand from 1834, published an account of a journey through the North Island. He noted more than a dozen plants that were of value to the Māori, and included an appendix, furnished by the botanist Allan Cunningham, containing Latin and Māori names of plants. He recorded that kauri gum was collected and sold to American traders at 'tuppence ha' penny' a pound.[15]

A year later (1843) Dr Ernest Dieffenbach's account of his travels as naturalist to the New Zealand Company during the years 1839-1842 was published. Dr Gerda Bell, who wrote on Dieffenbach's life, commented that

> he knew parts of New Zealand better than anyone else visiting the country during this period. In his conclusions and forecasts he was frequently ahead of his time. He anticipated topical issues of today such as ecology and the protection of the environment. Thus. . . he particularly deplored the ruinous felling of trees, an exploitation for which future generations would pay dearly because of erosion.[16]

In a chapter on botany, Dieffenbach noted various uses of plants; the Māori used seaweeds to contain oil and other fluids, a large fungus as tinder, leaves of nīkau to roof houses, heart leaves of nīkau as food, stems of supplejack to support thatch-work, cultivated taro as food, bundles of raupō to form walls and roofs of houses, root of raupō as an 'amylaceous food', stems of mangemange to tie raupō bundles, bracteae of *Freycinetia banksii* as food, and leaves of kawakawa as a substitute for tea. Also he listed eight valuable native timber trees. The use of kawakawa leaves is not a traditional Māori usage; probably it was a European practice when imported tea was unavailable.[17]

John Logan Campbell, the 'Father of Auckland', made the first attempts to export native plant products. He and his partner Brown bought the *Bolina*, a barque of 318 tons, in 1844, and loaded it with copper ore, manganese, dyewoods, wools, sarsaparilla, 'substantial quantities of wood and bark extracts', flax, kauri spars, sulphur, furniture wood, kauri deck planks, boat knees, kauri gum, tanning bark, and five cases of live plants for Kew Gardens. This was the first export cargo from Auckland to the United Kingdom. The *Southern Cross* newspaper of 14 December 1844 said that 'Messrs Brown and Campbell who are the spirited owners of this ship deserve the thanks of the whole community for their praiseworthy efforts', and went on to state that all the items were obtained within sight of Auckland, the Firth of Thames, or the Hokianga. The fact that Brown owned the *Southern Cross* may have helped this telescopic view. The pioneer speculation was not a success: the British Navy was in a period of retrenchment, and was not interested in spars or decking, which were eventually sold at a considerable loss. It was reported (Stone, 1982) that the result of the sale of the *Bolina* cargo, and three similar ventures, only covered costs.[18,19]

Reverend Richard Taylor joined the mission station at Paihia in 1839, and soon began recording the natural history of northern New Zealand. In 1847 he sent a list of 'vegetable productions' to the *N. Z. Church Almanac*, and this was reproduced in *The New Zealand Journal* for 25 March 1848. The *Journal* was published fortnightly in London, at a price of sixpence, with news for intending emigrants. Taylor listed twenty-two timbers, with their uses, measurements of

Reverend Richard Taylor, M.A., F.G.S., 1805-1873
From neg. B1154, Library, Auckland Institute and Museum

stiffness, strength, and toughness. Varieties of New Zealand flax are recorded in order of fineness, with brief notes on cultivation and preparation. This is followed by a table of six dyewoods, mako (*Aristotelia serrata*) (blue black) and whākou (*Ixerba brexioides*) (blue) being recommended for domestic purposes. The list of food plants is surprisingly long, with more than forty species.[20]

Taylor contributed a list of trees to *The New Zealand Journal* for 9 September 1848, with useful notes on the timber of tōtara, pūriri, rewarewa, pōhutukawa, rimu, and white pine, but with more information on the fruit of mataī, miro, rimu, hīnau, tutu, and karaka, and on the dye colours of hīnau and tānekaha barks, and the virtues of tawa bark tea. In the same year (1848) Taylor's first book was published, in which he listed Māori names of animals, birds, fish and plants. There are many errors in the book; those in the sections on plants seem to be typographical errors. The records of Māori and Pākehā plant remedies are listed in Brooker, Cambie and Cooper's *New Zealand Medicinal Plants*, and other items of interest are included in later chapters of this work. There is a severe criticism of the Māori dictionaries compiled by Taylor in Williams' preface to the 5th edition of *A Dictionary of the Maori Language*. Colenso, Taylor's rival in all matters Māori, botanical, and to do with the moa, said that Taylor knew no Māori; but Bagnall and Petersen, authors of a biography of Colenso, considered this to be an exaggeration.[21-25]

Taylor's second book, *Te Ika A Maui, or New Zealand and its Inhabitants*, was published in 1855 and reprinted with more information in 1870. In the chapter on botany, some sentences were taken from Dieffenbach, with little change and without acknowledgement, but other information seems to be original. For example, in commenting on the taste of Māori foods, he described the tender shoot of ti rākau (*Cordyline australis*) as 'rather bitter', the ripe fruits of kahikatea (*Dacrycarpus dacrydioides*) as 'sweet, but without flavour' (Plate 2), and the fleshy bracts of the flowers of kiekie (*Freycinetia banksii*) as 'very sweet, with a flavour not unlike a luscious pear'.[26]

Predictions are always of interest, and Taylor made several: he said of the fibre of New Zealand flax that:

> although it has gained much celebrity from the fine mats made of it by the natives, and also for the strength of its fibre in the manufacture of rope, it will, I am persuaded, soon be better known as a substitute for rags in the manufacture of paper, for which purpose it is admirably adapted.

This prophecy about flax may yet prove to be right. In the 1980s the technical staff of N.Z. Forest Products Ltd conducted large scale trials on growing the plant and making paper from it. The product was perfectly satisfactory, but not economically viable. Nevertheless, with the insatiable demand for paper, which is putting an increasing strain on our wood resources for the purpose, it could come into its own.[27]

Again, Taylor made a suggestion that several members of the *Epacris* family, being extremely ornamental and belonging to alpine regions, might be introduced successfully into English gardens. This prophecy has been fulfilled; not only the native snowberries but many other species of the alpine flora of New Zealand have proved worthwhile garden plants overseas. Today it seems a great pity that Taylor and Colenso could not have collaborated in their Māori and botanical studies.[26]

William Colenso, printer to the mission station at Paihia, arrived at the Bay of Islands from Sydney in December 1834 and spent the rest of his life in New Zealand. He died at Napier in 1899. He was encouraged to study plants by Allan Cunningham, Colonial Botanist of Sydney, when he visited Paihia, and by Dr Joseph Hooker when, as assistant surgeon and naturalist to the *Erebus*, he spent from August to November 1841 at the Bay of Islands. Both visitors made field trips with Colenso, whom Hooker considered to be the 'foremost New Zealand botanical explorer'.[10,25,28]

When Dr James Hector was appointed Commissioner of the N.Z. Exhibition, held at Dunedin in 1865, he chose Colenso to write essays 'On the Maori Races of New Zealand' and 'On the Botany of the North Island of New Zealand'. These two authoritative articles were included in the ten essays on aspects of local science which were published in the first volume of the *Transactions of the New Zealand Institute* in 1869. The wealth of information that Colenso recorded on economic plants is included elsewhere in this work where it is relevant.[29,30] (See the chapters on foods and beverages, land, marine and freshwater plants, potions and poisons, and others.)

In the Botany essay, Colenso suggested that a number of native plants, judging by their known natural affinities, might prove valuable in medicine. He supplied a long list of trees, shrubs and herbs, at that time unknown in European and Colonial gardens, which he thought worthy of a place in horticulture. He also listed a number of plant products which he thought might prove valuable in future: known but unused hardwoods, reputed hardwoods, and barks suitable for dyeing and tanning. He suggested that the leaves of the kiekie would make excellent hats, that the oils of the berries of the tītoki and from the leaves and bark of pukatea would prove 'serviceable', that the gum from the base of the leaves of N.Z. flax might be collected and exported, and that the leaves of seagrass might be useful for stuffing mattresses and making paper.[29]

In the essay on Māori Races, Colenso described two edible roots, kūmara and taro, a gourd-like fruit called hue, and the cloth plant known as aute. He gave a long list of wild edible substances, and shorter lists of plants used for Māori buildings and for making mats, baskets, sandals, nets, lines, canoe sails, calabashes, flutes, etc. Finally, he identified the hardwoods chosen by the Māori for weapons, digging spades, canoe paddles, fish and bird spears, and so on. From 1865 to 1895, Colenso published papers almost every year in the annual volumes of the *Transactions of the New Zealand Institute*. As well as contributions to botany and zoology, he wrote extensively on the Māori people, with further notes on the uses made in past times of the plants of New Zealand.[25,30]

John Buchanan, a printmaker and draughtsman, emigrated to New Zealand in 1852 and settled in Dunedin. He was appointed draughtsman and botanist to the Otago Geological Survey in 1862, and accompanied Dr Hector on several journeys to explore central and western parts of the Otago province. Hector invited Buchanan to write an account of the botany of Otago for the New Zealand Exhibition of 1865, and this was later reprinted in the first volume of the *Transactions of the New Zealand Institute*, published in 1869. In the essay, Buchanan assessed the value of different species of native grasses for pasture (see Chapter 7), and noted other useful Otago plants. He deplored the deterioration of repeatedly-burnt tussock pastures, and the exhaustion of tree mānuka firewood supplies near Dunedin.[31,32]

The reports of the jurors of the 1865 Exhibition contained a detailed account of flax exhibits, which came from Hawke's Bay, Wellington, Nelson, Canterbury, Otago, Southland and Tasmania. Auckland was not represented, possibly owing to the recent 'native rebellion'. The jurors provided an interesting discussion of the flax trade, with export figures for the years from 1828 to 1864. They noted the potential value for paper-making of N.Z. flax, cabbage tree, nīkau and kiekie leaves, native grasses, bark and softwood of the Mallow family (*Plagianthus* and *Hoheria*), and the floss or scales of *Astelia*. In listing timber exhibits, they remarked that native cork wood (*Entelea arborescens*) would make admirable fenders, linings for life-boats, and filling for life-belts. They described kauri gum and gave the value and quality of gum exported each year from 1853 to 1864. In addition they commented favourably on timber polished with rimu gum and noted that rimu gum varnish was available at twenty-one shillings per gallon.[33]

In 1867 Dr Joseph Hooker completed his *Handbook of the New Zealand Flora*, and in it, and in the earlier *Flora Novae-Zelandiae* there are brief notes on the uses of native plants. They were made from his own observations and from information recorded by Colenso, Dieffenbach, Bidwill, Taylor, and others. Mention must be made here of the Museum which the Hookers developed at Kew along with the Herbarium and the collections of living plants. In the Museum, founded in 1847, they gathered an outstanding collection of plant products from all parts of the world and especially from the British colonies.[10,28,34]

Thomas Kirk arrived in New Zealand in 1863, and soon became the leading resident botanist. Kirk served as Secretary of the Auckland Institute and Curator of the Auckland Museum from 1868 to 1873, and made several contributions to our knowledge of indigenous economic plants during those years. He spoke on N.Z. flax and other native plants as paper-making materials, listed plants best adapted for the reclamation of sand, and compiled notes on the native plants available for food, medicine, and other economic purposes. His most important work, *The Forest Flora of New Zealand*, published by the Government Printer, Wellington, in 1889, contained a large amount of information on the uses and value of native trees, shrubs, and timbers. This fine book has been the inspiration for several popular illustrated works such as *The Trees of New Zealand* by L. Cockayne and E. Phillips Turner (1928) and subsequent editions, *The Trees of New Zealand* by J. W. Matthews (1951), and *Trees and Shrubs of New Zealand* by A. L. Poole and N. M. Adams (1964 and subsequent editions). Many of Kirk's notes are included in later chapters.[35-41]

Leonard Cockayne settled in New Zealand in 1880, taught at a public school in Otago from 1881 to 1884, farmed for a few years at Styx in Canterbury, and then bought four acres of land near New Brighton and devoted his time to growing and studying native plants. He died at Wellington in 1934. He introduced to New Zealand the study of seedling forms, plant ecology and plant hybridization. Knowledge of economic native plants was advanced by his study of their role in the colonization and stabilization of sand dunes, his examination of the over-grazed areas of Central Otago, his report on the ecology and management of beech forests, his pioneer study of the palatability for sheep of tussock grassland and forest plants, and his classic little book on the cultivation of New Zealand plants.[42-45]

A contemporary of Cockayne was Elsdon Best, who lived with Māori people for many years before joining the staff of the Dominion Museum (now the National Museum), Wellington, and writing accounts of his field work. His *Maori Forest Lore* (1908), containing lists of the many native plants used by the Tuhoe people of Urewera, and *Maori Agriculture* (1925) on cultivated food plants, and his many other papers and books remain valuable reference works for anyone interested in ethnobotany.[46,47]

2.3 Botany Division, DSIR

In 1928 the New Zealand Government established a Plant Research Station at Palmerston North as a joint enterprise of the Department of Agriculture and the new Department of Scientific and Industrial Research. Here G. H. Cunningham and H. H. Allan carried out the services which later became the Plant Diseases Division and Botany Division of DSIR. Cunningham studied fungi; Allan worked on weeds and grasses; V. D. Zotov, who joined the staff in 1928, had various duties including the collection of flower pollens to compare with those of honey; C. E. Woodhead, who joined in 1931, conducted trials of rootstocks for fruit trees.

Staff appointed in later years worked on a variety of problems. New Zealand flax at the Moutoua Estate in the Manawatu, ragwort, hard fern, Nassella tussock, poisonous plants, seeds, foods of wildlife, aquatic plants, and depleted tussock grassland were all subjects of study. During the Second World War, 1939-1945, when many imports were unobtainable, officers of Botany Division began intensive studies of introduced and native plants to find substitutes. Native sow thistle (*Sonchus* sp.) was examined as a possible substitute for rubber, N.Z. flax and cabbage tree leaves for fibre, and sphagnum moss for surgical dressings. Botanical societies were asked to collect lycopodium powder for photographic flares.[48]

Studies of seaweeds, by Lucy Moore, led to a seaweed industry which, since the end of the Second World War, has continued to make agar from species of *Pterocladia* and small amounts of carrageenan from species of *Gigartina*. While the staff of the Botany Division sought substitutes for imports in short supply, the staff of Plant Diseases Division were extensively involved in the later war years with problems of disease control in the greatly expanded vegetable areas south of Auckland, and with such wartime difficulties as the rotting of canvas tents under tropical conditions.

In 1948 Allan retired as Director of Botany Division and was commissioned by DSIR to rewrite Cheeseman's *Manual of the New Zealand Flora* (1925), but soon began the preparation of a new *Flora of New Zealand*. Other members of the staff of the Division who have assisted with the *Flora* are Lucy Moore, Elizabeth Edgar, Arthur Healy, David Galloway and others; work continues on further volumes. Urgent ecological and conservation problems have demanded staff time and expertise, and the study of the flora for new products has had little attention at Botany Division since the Second World War. Henry Connor has examined poisonous plants and the chemistry of native grasses, Murray Parsons has looked at the potential of agar and carrageen seaweeds, and Neville Moar has studied the pollen analysis of honey, but generally this field has become the prerogative of the Chemistry Departments of the Universities, the DSIR (new compounds), the Forest Research Institute (forest products) and industry.[48-51]

Beginning in the 1985/86 financial year, the funding of the Botany Division was progressively cut by 27% and staff who left were not replaced. One result was a sharp reduction in the activities of the Division.[51] In September 1986, the Government policy of recovering costs for services, known as 'the User Pays policy', was adopted, and a charge of $20 was introduced for each plant identification. Requests for plant names and information fell from 1177 in 1985 to 204 in 1987. The loss of reference specimens and data about them forced the Division to abandon 'User Pays' from August 1989, except where clients obviously benefit commercially from identifications.[52]

In 1987 the Division started several projects of an economic nature. The ethnobotanical project included propagation studies to provide a saleable recipe for growing pīngao, and the setting up of a national collection of flaxes. The poisonous plant project was a new pharmacological survey, with Botany Division staff supplying the plant material and keeping records of it, and Chemistry Division staff making analyses. A third project included the evaluation of attractive native plants with a view to their sale to nurseries — especially Marlborough daisies, tea-trees, hebes, kāka-beak, and native brooms from the experimental gardens of Botany Division. Other plans were for a survey of vegetative reproduction in monocotyledon cultivars, a study of fructose production from cabbage trees and of essential oil production from tea-trees.[53]

However, as a further economy, DSIR decided to cut the number of its divisions and, at 31 March 1990, disestablished Botany Division. From 1 April, a new and larger Land Resources Division was set up, in which remnants of Botany Division survive as Land Resources Botany Institute. Hopefully, the economic projects recently started will be completed.[54] In 1991, planning began for the disestablishment of DSIR, and the setting up of Crown Research Institutes (CRIs).

2.4 Fruit and Trees

Plant Physiology Division of DSIR, situated at Palmerston North, studies the impact of environment on agricultural and horticultural crops; environmental, plant cell, and crop physiology; environmental physics; climate laboratory experiments; and tree and shrub improvement (mainly of introduced poplars, willows, legumes, and grasses). Native plant studies have included the mucilage from whau leaves (1980), Climate Laboratory screening of native plants for frost hardiness and frost tolerance (1982-1987), and growth chamber experiments to determine *Hebe* flowering criteria (1989).[55-58]

A Horticultural and Processing Division of DSIR, formed in 1980 at Mt Albert Research Centre, also worked mainly on introduced plants, especially kiwifruit; however, during its short life it successfully launched a liqueur based on tītoki berries. In 1989, Plant Physiology and Horticulture and Processing Divisions were combined under the new, more 'public friendly' name of 'Fruit and Trees', abbreviated to FAT. As DSIR divisional budgets continue to shrink, it seems an unfortunate acronym.

2.5 Other Contributors

The gardeners' role in utilizing New Zealand native plants in horticulture is described in Chapters 3 and 4, the foresters' logging of timber trees in Chapter 5, and the chemists' search for new compounds in Chapters 9 to 12. In *New Zealand Medicinal Plants*, 1987, we described very briefly some past attempts by industry to use native plant products. Some of these stories are expanded in appropriate chapters of this book; others are omitted as they have been told fully elsewhere by experts in the fields concerned; for example, G. G. Thornton has dealt with paper-making and the New Zealand flax industry in his well-illustrated book, *New Zealand's Industrial Heritage*, 1982.[59]

Three of the saddest items in the story of the economic plants of New Zealand have been the difficulties of New Zealand firms to establish an export trade in manool, the collapse of the Waitara *Solanum* venture, and the inability of the Kaurex plant to produce resin profitably. We may note here that successive New Zealand governments have fallen far behind some other countries in encouraging scientific research and development by offering tax rebates and other financial help.

1. Hooker, J.D. (ed.) *Journal of the Rt. Hon. Sir Joseph Banks Bart. K.B., P.R.S., during Captain Cook's First Voyage in H.M.S. Endeavour in 1768-1771, etc.* London (Macmillan) 1896 466p.
2. Beaglehole, J.C. (ed.) *The Endeavour Journal of Joseph Banks 1768-1771* Sydney (Trustees of the Public Library of New South Wales in association with Angus & Robertson) 1962 2 vols. Vol. 2, pp. 8, 10, etc.
3. Dunmore, J. (transl. and ed.) *The Expedition of the St. Jean Baptist to the Pacific,*

1769-1770, from the Journals of Jean de Surville and Guillaume Labé London (The Hakluyt Society) 1981 310p.

4. McNab, R. *Historical Records of New Zealand* Wellington, (Government Printer) 1914 2 vols., reprinted 1973. Vol. 2, p. 296 Extracts from the Journal of P. de L'Horne, 1769-1770.

5. Cook, J. *A Voyage towards the South Pole and Round the World. Performed in His Majesty's Ships The Resolution and Adventure, in the Years 1772, 1773, 1774, 1775* London (Strahan & Cadell) 1777 2 vols.

6. Forster, G. *A Voyage round the World in his Britannic Majesty's Sloop Resolution . . .* London (White) 1777 2 vols.

7. Forster, G. *De Plantis Esculentis Insularum Oceani Australis Commentatio Botanica* Berlin (Haude & Spener) 1786 80p.

8. Hatch, E.D. (1986) 41 (2) *Auckland Bot. Soc. Newsletter* 70.

9. Britten, J. (1916) 54 *J. Bot.* 345, (1917) 55 *J. Bot.* 54.

10. Hooker, J.D. *Handbook of the New Zealand Flora* London (Lovell Reeve) 1864 Vol. 1, 1867 Vol. 2.

11. Rienits, R., Rienits, T., *The Voyages of Captain Cook* London (Paul Hamlyn) 1968 157p.

12. Cheeseman, T.F. *Manual of the New Zealand Flora* Wellington (Government Printer) 1925, 2nd ed., revised by the author, edited by Oliver, W.R.B. 1163p.

13. McNab, R. *Murihiku and the Southern Islands* Invercargill (W. Smith) 1907, 377p, reprinted by Wilson & Horton, n.d.

14. Yate, W. *An Account of New Zealand* London (Sealey & Burnside) 1835, 310p.

15. Wade, W.R. *A Journey in the Northern Island of New Zealand* Hobart Town (G. Rolwegan) 1842 206p., reprinted by Capper Press, 1977.

16. Bell, G. *Ernest Dieffenbach, Rebel and Humanist* Palmerston North (Dunmore Press) 1976 165p.

17. Dieffenbach, E. *Travels in New Zealand* London (J. Murray) 1843 2 vols., reprinted by Capper Press, 1974.

18. *Southern Cross*, 14 December 1844. Exports by Brown & Campbell.

19. Stone, R.C.J. *The Young Logan Campbell* Auckland (Oxford University Press) 1982 287p.

20. [Taylor, R.] (1848) 217 *New Zealand Journal, London* 68 'Vegetable Productions of New Zealand'. Extracted from the *New Zealand Church Almanack* for 1847.

21. [Taylor, R.] (1848) 229 *New Zealand Journal, London* 229 'Names and Properties of the Principal New Zealand Trees'. From the *Annals of the Diocese of New Zealand*.

22. Taylor, R. *A Leaf from the Natural History of New Zealand* Wellington (Stokes) and Auckland (Williamson) 1848 102p. 2nd ed. Auckland (Chapman) 1870 120p.

23. Brooker, S.G., Cambie, R.C., Cooper, R.C. *New Zealand Medicinal Plants* 2nd ed. Auckland (Heineman) 1981 117p., 3rd ed. 1987 268p.

24. Williams, H.W. *A Dictionary of the Maori Language* 5th ed. Wellington (Government Printer) 1917, 6th ed. 1957, 7th ed. 1971.

25. Bagnall, A.G., Petersen, G.C. *William Colenso* Wellington (Reed) 1948 494p.

26. Taylor, R. *Te Ika A Maui or New Zealand and Its Inhabitants* London (Wertheim & Macintosh) 1855 490p., 2nd ed. London (Macintosh) and Wanganui (Jones) 1870 713p., reprinted by Reed, 1974.

27. Mackney, A.W. (1985) Address to N.Z. Institute of Chemistry, Auckland.

28. Hooker, J.D. *The Botany of the Antarctic Voyage of H.M. Discovery Ships Erebus and Terror in the Years 1839-1843* London (Reeve) 6 parts in 3 vols. Vol.

1, 2 parts: *Flora Antarctica* 1844, 1847. Vol 2, 2 parts: *Flora Novae Zelandiae* 1852, 1855. Vol. 3, 2 parts: *Flora Tasmaniae* 1855, 1860.

29. Colenso, W. (1869, 2nd ed. 1875) 1 *Trans. N.Z. Inst.* 233.
30. Colenso, W. (1869, 2nd ed. 1875) 1 *Trans. N.Z. Inst.* 339.
31. Buchanan, J. (1869, 2nd ed. 1875) 1 *Trans. N.Z. Inst.* 181.
32. Hoare, M.E., Bell, I.G. (eds) *In Search of New Zealand's Scientific Heritage* Wellington (The Royal Society of New Zealand Bull. 21) 1984.
33. *New Zealand Exhibition 1865, Reports and Awards of the Jurors* Dunedin (Printed for the Commissioners by Mills, Dick & Co.) 1866.
34. Hepper, F.N. (ed.) *Royal Botanic Gardens, Kew: Gardens for Science and Pleasure* London (H.M. Stationery Office) 1982 195p.
35. Kirk, T. 'Plants Indigenous to New Zealand available for Food, Medicine, and other Economic Purposes' revised by T. Kirk. Typed manuscript in Auckland Institute and Museum Library n.d. *c* 1870.
36. Kirk, T. (1870) 2 *Trans. N.Z. Inst.* 102.
37. Kirk, T. (1871) 3 *Proc. N.Z. Inst.* 48.
38. Kirk, T. (1872) 4 *Trans. N.Z. Inst.* 267.
39. Kirk, T. (1872) 4 *Trans. N.Z. Inst.* 295.
40. Kirk, T. (1874) 6 *Trans. N.Z. Inst.* 45, 55.
41. Kirk, T. *The Forest Flora of New Zealand* Wellington (Government Printer) 1889 142pl., 345p.
42. Cockayne, L. *Report on the Sand Dunes of New Zealand. The Geology and Botany with their Economic Bearing* Wellington (Government Printer) 1909 30p.
43. Cockayne, L. *Report on the Dune-Areas of New Zealand, their Geology, Botany and Reclamation* Wellington (Government Printer) 1911 74p. Published with Cockayne L. 1909 as Appendices of the Journals of the House of Representatives, C13 1909: 1-30 and C13 1911: 1-74.
44. Cockayne, L. *The Cultivation of New Zealand Plants* Christchurch (Whitcombe & Tombs) 1924 139p.
45. Cockayne, L. *The Vegetation of New Zealand* Leipzig (W. Engelmann) 1928 2nd ed., 456p. (Engler & Drude, *Vegetation der Erde* Part XIV).
46. Best, E. (1908) 40 *Trans. N.Z. Inst.* 185, (1909) 41 ibid. 231, (1910) 42 ibid. 433.
47. Best, E. *Maori Agriculture* Wellington (Whitcombe & Tombs) 1925 (Dominion Museum Bull. 9) 172p.
48. Atkinson, J.D. *DSIR's First Fifty Years* Wellington (N.Z. DSIR) 1976 220p.
49. Allan, H.H. *Flora of New Zealand Volume 1* Wellington (Government Printer) 1961 1085p.
50. New Zealand Department of Scientific and Industrial Research Annual Reports, 1948-1988.
51. New Zealand DSIR Botany Division 1985-87 Triennial Report. Christchurch (Caxton Press) 1989, 106p, and earlier issues.
52. Parsons, M. (1989) 127 *Botany Division Newsletter* 11.
53. Molloy, B. (1990) 130 *Botany Division Newsletter* 3.
54. Connor, H.E. (1990) 20 *New Zealand Botanical Society Newsletter* 14.
55. Haslemore, R.M., Roughan, P.G. (1980) DSIR Plant Physiology Division 1980 Report 26.
56. Wardle, P. (1982) 13 *Climate Laboratory Newsletter* 19.
57. Wardle, P. (1983) 14 *Climate Laboratory Newsletter* 18.
58. Warrington, I. (1989) 27 *Biological Industries Group Newsletter* 3.
59. Thornton, G.G. *New Zealand's Industrial Heritage* Wellington (Reed) 1982 194p.

3

Native Plants in Gardens Overseas

3.1 Seeds from Cook's First Voyage

Mr Joseph Banks, Dr Solander, and their assistants collected seeds of New Zealand plants during their visits ashore at Poverty Bay, Anaura Bay, Tolaga Bay, Mercury Bay, River Thames (now the Waihou River), and the Bay of Islands in the North Island, and Queen Charlotte Sound and Admiralty Bay in the South Island. Live plants on board ship had no chance of surviving the voyage home through the tropics so seeds were collected, dried, and wrapped in paper or wax.

The seeds reached England in July 1771. Shortly afterwards, Princess Augusta died. George III joined her estate to his property at Kew, and invited Mr Banks to direct the two gardens 'in a scientific manner'. Consequently, most of the seeds were grown at the Royal Gardens, Kew, although some were raised at the Chelsea Physic Garden, at Bank's country residence at Spring Grove, and at other gardens and nurseries.[1-3]

HORTUS KEWENSIS
Banks, Solander, and Carl von Linné, son of the immortal Linnaeus, spent four days a week through the summer of 1781 cataloguing the plants at Kew. After Solander's death in 1782, Jonas Dryander, another Swedish botanist and pupil of Linnaeus, completed the catalogue, in which 5500 species of plants were named, diagnosed, and classified according to the 'Sexual System' of Linnaeus.

Hortus Kewensis or a Catalogue of the Plants Cultivated in the Royal Botanic Garden at Kew was published in three volumes in 1789 with William Aiton, Head Gardener to His Majesty, as the author. It was not until the second edition by Aiton's son was published in 1810-1813 that Dryander's contribution was acknowledged.[4-6]

New Zealand plants grown at Kew and listed in 1789 were:
whorl'd-flowered Haloragis, *Haloragis cercodia* (now *H. erecta*); fig marygold,

Mesembryanthemum australe (now *Disphyma australe*); sweet-scented Philadelphus, *Philadelphus aromaticus*; narrow-leav'd Philadelphus, *Philadelphus scoparius* var. *linifolius*; myrtle-leav'd Philadelphus, *Philadelphus scoparius* var. *myrtifolius*, (all three now treated as *Leptospermum scoparium*); cut-leav'd nightshade, *Solanum laciniatum*; small-leav'd shrubby Sophora, *Sophora microphylla*; winged-podded Sophora, *Sophora tetraptera*; horn'd Tetragonia, *Tetragonia expansa* (now *T. tetragonioides*).

Aiton gave the source of each species as 'New Zealand, Sir Joseph Banks Bart., 1772', except for *Mesembryanthemum australe*, for which the date of introduction was '1773'.[5,6]

CHELSEA PHYSIC GARDEN
From seed taken back to England by Mr Joseph Banks in 1771, plants of *Sophora tetraptera* were raised successfully and Mr Forsyth planted one at Chelsea in 1774. It flowered in 1779, and John Miller drew it for his *Icones Novae* of 1780. Dr W. Pitcairn planted another seedling in his garden at Islington, and it also flowered in 1779. William Curtis published a lovely hand-coloured engraving of *Sophora tetraptera* in 1791, with the comment

> A finer sight can scarcely be imagined than a tree of this sort, extending to a great breadth on a wall with a western aspect, in the Apothecaries Garden at Chelsea . . . and which at this moment (April 28, 1791) is thickly covered with large pendulous branches of yellow . . .[7,8]

A further five illustrations of New Zealand's *Sophora tetraptera* were published before 1800.[9]

3.2 Seeds from Cook's second voyage

Any seeds collected in New Zealand during Cook's second voyage would have reached England in July 1774 on the *Adventure*, and in July 1775 on the *Resolution*. Dr Solander wrote to Sir Joseph Banks on 22 August 1775, advising that 'Captain Cook has sent all his curiosities to my apartment at the [British] Museum'. There is no mention of any seeds.[10] Dr Johann Forster and his son George collected New Zealand plants at Dusky Sound and Queen Charlotte Sound during the second voyage; again, there is no mention of any seeds being brought back to grow.[11,12] However, John Hope, Professor of Botany at Edinburgh, wrote to Sir Joseph Banks on 16 August 1779, describing a beautiful evergreen shrub with orange-coloured flowers, grown from seed sent to him by William Anderson, who sailed on Cook's second and third voyages, and asking whether Banks had also brought it home from the

large-leaved kōwhai (*Sophora tetraptera*) **Fabaceae**
From C. Loddiges & Sons *Botanical Cabinet* (1826)

first voyage. Presumably the seed was from the second voyage, as Anderson died at sea in 1778 before the completion of the third voyage.[13,14]

In February 1780, Professor J.A. Murray of Goettingen named, described, and illustrated *Cercodia erecta* Banks, now known as *Haloragis erecta*. Murray grew the plant from seed which he obtained in 1778 from Professor Jacob R. Spielman of Strasbourg, a correspondent of Sir Joseph Banks. However, Murray obtained the name '*Cercodia erecta* Banks' from J.R. Forster and it is possible that the seed came originally from the same source.[15]

3.3 Seeds from Cook's third voyage

David Nelson, a young gardener, is credited with collecting 386 kinds of seeds from the South Seas, including about forty-seven species from Queen Charlotte Sound in February 1777. Possibly the Forsters' *Characteres Generum Plantarum* was available before the *Resolution* and *Discovery* sailed from England in 1776, but the plant names used by Nelson for the New Zealand seeds seem to be those of Solander's unpublished 'Primitiae Florae Novae Zelandiae', and are not reproduced here. However, the following common coastal plants can be recognized in the list:[16-18] *Acaena, Aciphylla squarrosa, Arthropodium cirratum, Astelia, Carex, Clematis, Convolvulus, Coprosma, Cordyline australis, Coriaria, Cyathodes, Epilobium pallidiflorum, Euphorbia glauca, Gahnia, Griselinia, Haloragis, Lepidium oleraceum, Libertia, Linum monogynum var. grandiflorum, Macropiper, Melicope ternata, Melicytus, Metrosideros, Myoporum laetum, Myrsine, Olearia, Phormium, Pittosporum, Pseudopanax,* and *Ripogonum scandens.*

The seeds reached England in October 1780 and Dr Solander sent surplus seeds to the following people: Aiton, Fothergill, Pitcairne, Lee, Forsyth, Malcolm, Thouin, Lord Sandwich for Mrs Gawler, Lord Bute, Count d'Artois, Mrs Banks, Lieut. Roberts, Dr Russell, Sir Wm Musgrave for Sir John Frederick, Lord Tankerville, Marquis of Rockingham, and M. de Wevelsinhoven. He also sent seed to the following places, presumably to the botanical gardens in these towns: Edinburgh, Cambridge, Petersburg, Goettingen, Lund, Amsterdam, Paris (Jardin du Roi), Florence, Vienna, Padova, Milano, Louvain.

Most of the recipients can be identified without difficulty. The letters of many of them have been preserved and are summarized in W.R. Dawson's *The Banks Letters*, 1958. 'Aiton' was William Aiton, Head Gardener to His Majesty at the Royal Botanic Garden, Kew. 'Fothergill' was John Fothergill M.D., F.R.S., who had a 'botanic garden' at Upton. 'Pitcairne' was William Pitcairn M.D., who had a 'botanical garden' of five acres in Upper Street, Islington. 'Lee' was James Lee, nurseryman at the Vineyard, Hammersmith. 'Forsyth' was William Forsyth, Curator of the Chelsea Physic Garden. 'Malcolm' was William Malcolm of the Kennington Nursery near Lambeth. 'Thouin' was Head Gardener and Professeur de Culture at the Jardin du Roi, Paris, and so on.

If and when the seeds grew, some of the recipients sent the plants to the Banksian Herbarium for identification. Others named, described, and illustrated them, sometimes with little regard for their existing names. In the Day Book of the Banksian Herbarium for the period 1777 to 1796, Daniel Solander and Jonas Dryander listed specimens sent for identification and among the senders were the following recipients of seeds: the Earl of Bute, Dr J. Fothergill, J. Lee, W. Malcolm, Dr W. Pitcairn, Lord Tankerville, Kew Gardens, the Apothecaries' Garden at Chelsea, and Edinburgh Botanical Garden.

3.4 Wardian cases

The first attempt to take New Zealand plants to England in a plant case seems to have been made in 1791, at the instigation of Sir Joseph Banks. The *Discovery*, under the command of Captain George Vancouver, carried a plant case when she sailed from England in 1790. From 2 to 24 November 1791, Archibald Menzies, surgeon and naturalist to the expedition, collected at Dusky Sound and, in his journal for the 18th, recorded taking live plants on board. Among them was the New Zealand flax plant, which he hoped to carry in the frame on board to His Majesty's Gardens.*

Towards the end of the voyage, the Commander requested that all journals and papers be handed to him in accordance with Admiralty instructions. Menzies refused, and was placed under arrest on 28 July 1795. He soon saw reason, conformed, and was released. However, the live plants in the case were neglected during his imprisonment, and died. In January 1793, when Lieut. Broughton was sent home with despatches, Menzies forwarded a box of seeds for Sir Joseph Banks.[13,34]

Nathaniel Bagshaw Ward, a medical practitioner of Whitechapel and a friend of Anderson at the Chelsea Physic Garden, the Hookers at Kew, and Lindley at the Horticultural Society, knew of the difficulties of importing live plants, and devised the Wardian case in 1831. He built a little, portable greenhouse, and made it water-tight with paint and putty, advising users to keep it closed and to water it once every five or six weeks. In about 1834 Loddiges sent a trial shipment of plants to Australia in Wardian cases and when word was received that the plants had arrived in good condition, the cases came into general use.[35]

As early as 1850, Sir William Hooker of Kew exchanged plants with Sir George Grey in New Zealand, but losses were heavy as the Wardian cases passed through the tropics. After Sir William's death in 1865, his son Dr (later Sir) Joseph Hooker continued the exchanges with similar losses.[36]

Wardian cases were disliked by some masters and crew, and other means were tried; for example, in 1850, William Brown of Brown and Campbell, the Auckland merchants, brought roots and seeds from England packed in a box and a jar filled with sawdust. William Wilson, a pioneer nurseryman of Christchurch, used airtight casks about 1858 to import seeds packed in canvas bags, which were wedged in a cask with dry chaff. However, the success of casks depended on the cooper; if the cask was well made, there was no need to use bags to hold the seed, but if the cooper used green timber, the staves shrank and seed was lost or damaged by sea water.[37-39]

Wellington Botanical Garden also tried barrels to bring out roots and cuttings from Kew, but losses from rot were heavy. The problem was not solved until it became possible to send plants in plastic bags by air.[36]

---

* Banks designed a 'plant cabin' for the safe conveyance of the King's plants as he could see no excuse for the fact that 'sometimes when the Captain's turtle comes home safe . . . the King's plants should die'. This note is from Gilbert, L.A. (1966)1 (1) *Rec. Australian Acad. Sci* n.p. quoting Banks to Wm Devayne, 6 June 1794, Brabourne Papers, Vol. 18, p.51, Mitchell Library of N.S.W.

3.5 Early plant names

A flora manuscript entitled 'Primitiae Florae Novae Zelandiae' was compiled by Daniel Solander during and after Cook's first visit to New Zealand, and contained scientific names, some Māori names, and descriptions in Latin of about 360 species of coastal plants. It was part of a much larger project on the botany of the voyage. Banks employed leading British artists to prepare 700 folio line engravings, of which about 200 were of New Zealand plants, from the drawings left by Sydney Parkinson and the herbarium specimens at Soho Square. Publication of the work was delayed by Banks' unsuccessful plans to accompany Cook on the second voyage, and the project seems to have been abandoned due to Solander's death in 1782 and Banks' increasing involvement as President with the affairs of the Royal Society.[16,18,40]

J.R. Forster and his son George wrote three works in which some New Zealand plants were named. The first, *Characteres Generum Plantarum*, contained brief descriptions of thirty-seven genera and 134 species of New Zealand plants along with some drawings of them; the second, *De Plantis Esculentis* of 1786 had descriptions of about fourteen edible plants found in this country; the third, *Florulae Insularum australium Prodromus* also of 1786, had brief descriptions, apparently by Anders Sparrman, of 141 New Zealand species, and twenty-three names without descriptions which served little purpose.[17,41,42]

Johann Sebastian Mueller (1715-1780), a German botanical artist, settled in London in 1744 and became known as John Miller. In his *Icones Novae* of 1780, there are two magnificent engravings of New Zealand plants. Tab. 1 is of the New Zealand kōwhai (*Sophora tetraptera*), which flowered at Chelsea and in Dr Pitcairn's garden in 1779. Tab. 2 is of New Zealand flax (*Phormium tenax*). It is probable that Miller drew the plant from a drawing or dried specimen in the Banksian Herbarium. The seeds of flax that Banks and Solander took to England in 1771 did not grow, and they were introduced again in 1788 or 1789 from Norfolk Island by way of Sydney.[6,7,43]

Another New Zealand plant to be named in the early 1780s was *Demidovia tetragonioides* (now known as New Zealand spinach *Tetragonia tetragonioides*). It was grown in the Moscow garden of Count Paul Demidoff from seed sent from Vienna by Baron N.J. Jacquin. Peter Simon Pallas F.R.S. of St. Petersburg, now Leningrad, catalogued the plants of the Count's Moscow garden in 1781, and his drawing of this plant is clear and accurate.[21]

New Zealand spinach was grown elsewhere in Europe in the 1780s. At Goettingen, it was named *Tetragonia expansa* by J.A. Murray in 1785, and at Pavia it received the same name from J.A. Scopoli in 1786. However, the information accompanying the seeds soon became garbled; for example, Pallas noted that the plant was probably of American origin.[23,27,21]

It was probably from the Botanic Garden of the University at Goettingen that the New Zealand tea plant (*Leptospermum scoparium*) was taken to Zurich as an indoor plant. There, in 1790, J.J. Roemer and P. Usteri named it *Philadelphus floribundus*. In 1796, H.A. Schrader and J.C. Wendland of Goettingen named and illustrated two new forms as *Melaleuca scoparia* var. *diosmatifolia* and var. *myrtifolia*. Meantime, in Britain, nurserymen were beginning to select, name and sell a number of forms (see below).[25,26]

Many botanists came from Europe to use the library and collections of Sir Joseph Banks at Soho Square, among them Joseph Gaertner F.R.S. (1732-1791) of Leiden, whose *De Fructibus et Seminibus Plantarum*, published in two volumes in 1788 and 1791, contained information about a number of New Zealand seeds and fruits collected on Cook's voyages.[45]

We have not dealt with other European botanists who named New Zealand plants at this time, for example, F.K. Medikus 1784, Charles L'Heritier de Brutelle 1791, and M. Vahl 1794, but enough has been given to show that after the distribution of seeds by Dr Solander in 1780 and the opening of the collections to visitors by Sir Joseph Banks, a number of authors named and renamed New Zealand plants.

The Napoleonic Wars hindered the entry to Britain and Europe of much more New Zealand plant material, and undue attention seems to have been given to the first introductions; for example, Murray and Jacquin published magnificent illustrations of the weedy *Haloragis erecta*, which was also drawn by the British artists employed by Sir Joseph Banks. However, Scopoli's 1786 drawing of New Zealand spinach is a work of art,[27] and Jacquin's 1798 plate of kōwhai (*Sophora microphylla*), flowering in the Schonbrunn Gardens, is superb (even in poor quality photocopy).[30]

3.6 Early British nursery catalogues

Leptospermum scoparium was offered for sale as 'Philadelphus aromaticus, 7s 6d' in a catalogue issued by Robert Anderson of Edinburgh sometime between 1776 and 1779. William Malcolm, 'Nurseryman and Seedman, Near Kennington Turnpike, Surry' (sic), offered four different kinds in his 1778 list of greenhouse plants:

Philadelphus Aromaticus Latifolia	Broad-leaved Aromatic	New
Philadelphus Angustifolia	Narrow-leaved	Zealand
Philadelphus Hirsutus	Hairy-leaved	Tea or
Philadelphus Lucidus	Shining-leaved	Myrtle

Dr Solander listed 'Malcolm' in the 1780 seed list; he must have had seed from the 1772 distribution in order to sell New Zealand tea plants in 1778.

By 1782, John Galpine of Blandford Forum in Dorset had 'Philadelphus aromaticus' in his catalogue of plants for sale. New Zealand tea-tree and possibly its Australian relatives must have been popular, for Conrad Loddiges, the Hackney nurseryman, offered *Leptospermum scoparium* and three other 'species' in his 1783 catalogue, eight named kinds in 1804, and ten in 1811, all classified as greenhouse plants.

In Adam Neal's 1779 catalogue of the plants in the garden of John Blackburne Esq. of Orford in Lancashire '*Philadelphus aromaticus*' and '*P. arom. fol. hirsutus*' were listed as greenhouse shrubs.

Both *Sophora tetraptera* and *S. microphylla* were offered for sale in 1783 by Loddiges, and Grimwood, another London nurseryman, had them also. Loddiges and Grimwood offered *Mesembryanthemum australe* (now *Disphyma australe*) in 1783, and the plant was listed by Loddiges from 1804 to 1818. Loddiges offered New Zealand flax for sale in each catalogue from 1804 to 1813; it was available to Irish gardeners at the same time or possibly earlier, and G. Don of Forfar in Scotland had it for sale in 1813. *Haloragis erecta* was also offered for sale in 1814 and 1818 but, being a weedy species with insignificant flowers, it did not become popular.[46-48]

3.7 England

FLOWER PRINTS

From the *Botanical Magazine*, the world's oldest botanical periodical, and from its numerous shorter-lived rivals, we can trace the introduction of many New Zealand plants into Britain. Mention has been made already of the successful raising of kōwhai at Kew, Chelsea and Islington. The first flowering of *Sophora tetraptera* in 1791 was recorded in Curtis's *Botanical Magazine* with a lovely hand-coloured engraving by Sydenham Edwards and a page of notes on the source of the plant and its cultivation.[8]

In 1796, William Curtis published the hand-coloured drawing of cut-leaved nightshade (*Solanum laciniatum*), noting that it flowered most of the summer, ripened fruit in the open air, and was easily increased by cuttings and seed. The berries when ripe 'are the size of a small plum, and of a yellowish-green hue; their pulp is sweet, in some degree resembling that of a fig; whether it be so innocent we do not take it on us to assert'. Apparently the first plants grown in Hungary in the early 1960s for the commercial production of solasodine were descendants of the Banks and Solander introduction.[49]

Flowering specimens of New Zealand tea (*Leptospermum scoparium*), sent by Mr Donn from the botanic garden at Cambridge in May and by Mr Milne from Fonthill in June, were illustrated by H.C. Andrews in the *Botanist's Repository* for 1812. He wrote that several varieties of it were in cultivation, and that it required to be kept in the greenhouse or conservatory. Andrews was an English botanical painter and engraver.[50]

New Zealand spinach (*Tetragonia expansa*, now *T. tetragonioides*) was figured by Samuel Curtis in the *Botanical Magazine* for 1822, the flowering specimen having been sent in by John Walker Esq. of St. Arno's Grove.[51]

Richard Davis, a missionary farmer from Waimate North and Rev. John Colman of Ryde, Isle of Wight, are credited with distributing seeds of the rare parrot's bill (*Clianthus puniceus*) in 1831 (Plate 2). Colman flowered a greenhouse plant and Mr L. Gower of Titsey Place near Godstow flowered it outdoors in May 1834. It figured in the *Transactions of the Horticultural Society of London* in 1834, and in *Edwards's Botanical Register* in 1836.[52,53]

The superb blue-flowered koru (*Colensoa physaloides*, now *Pratia physaloides*) was raised at Kew from seed sent by Dr James Hector in 1883, and flowered in a cool frame in 1885. It was illustrated in *Curtis' Botanical Magazine* for 1886, but it is not sufficiently hardy to attain the success it merits.[54]

The *Botanical Magazine* is still published, but since the end of the Second World War, hand-coloured plates have had to be replaced by machine printing and, in 1984, it underwent a name change to *Kew Magazine*. For a fuller account of flower prints, readers should refer to *Early New Zealand Botanical Art* by F. Bruce Sampson, published by Reed Methuen in 1985.[55]

KEW

John Smith of Kew, writing in 1880, maintained that credit for the first great importation of New Zealand plants after the Napoleonic Wars must be given to Sir William Symonds, Surveyor to the Royal Navy, who instructed the Commander and officers of *H.M.S. Buffalo* to procure young plants of the principal trees of New Zealand. *H.M.S. Buffalo*, after discharging convicts in Australia, visited New Zealand for return cargoes of kauri spars in 1833-1834, 1837-1838 and 1840 (when she was wrecked on Buffalo Beach, Whitianga). The three Wardian cases which arrived at Kew in good condition at the end of 1838, contained young plants of kauri (*Agathis australis*), rimu (*Dacrydium cupressinum*), tōtara (*Podocarpus totara*), pūriri (*Vitex lucens*), tānekaha (*Phyllocladus trichomanoides*), and many others.[56,118]

One of the first New Zealand gardeners to ship live plants to England was John Edgerley (*c* 1814-1849) who worked for Lieutenant Thomas McDonnell R.N. at Horeke, Hokianga, from *c* 1835 to 1841, and as a nurseryman at Newmarket, Auckland, from 1843 until his death in 1849. While at Hokianga, Edgerley collected live plants, pickled cones, and dried herbarium specimens for Aylmer Bourke Lambert F.R.S. (1761-1842), who apparently succeeded his friend, Sir Joseph Banks, as unofficial supervisor of the Royal Gardens, Kew. Unfortunately, Lambert was an old man; his affairs were in a state of confusion, and Edgerley received little credit for his plant collections.

In 1839, Lieutenant McDonnell took some boxes of New Zealand plants from the Hokianga to Mr W.T. Aiton, who had succeeded his father as Royal Gardener at Kew, and in 1842 Edgerley took further boxes to England when Sir William Hooker and Mr John Smith had charge of Kew Gardens. According to Smith's records, Edgerley delivered to Kew live plants of *Dicksonia squarrosa*, *Asplenium oblongifolium* (then known as *A. lucidum*), and *Hebe* (then known as *Veronica*). Edgerley sold some live plants to Mr Knight's Exotic Nursery at Fulham, and among them was the beautiful magenta-flowered napuka (*Hebe speciosa*), which was raised by Mr Knight and made available to Kew. It was illustrated in the *Botanical Magazine* for 1844 as the Showy-flowered speedwell (*Veronica speciosa*), with the information that it was too rare to risk outdoors in an English winter but thrived well in a cool greenhouse at Kew Gardens. It is one of the parents of a number of beautiful hybrids. Edgerley also sent or took live plants from Hokianga to Arley Hall, one of the seats of the Earl of Mountnorris, and McDonnell mentioned that they had shipped plants to the Duke of Bedford.[57,58]

About this time, the miniature tree fern (*Blechnum fraseri*) and many other plants were received at Kew from Rev. W. Colenso, but we have not found a list. A number of New Zealand species were sent to Kew by the Royal Botanic Gardens, Sydney, according to the 'List of Plants and Seeds Distributed' 1832-1847, and 1852-1869; for example, *Pittosporum eugenioides* was forwarded with other plants by the *Poisian* (Captain Mollard) on 14 April 1845.[59]

During the next 100 years, many New Zealand plants were introduced to Kew, flowered there and in other British gardens, were perhaps exhibited at the Royal Horticultural Society and illustrated in the *Botanical Magazine* or a similar horticultural journal. By 1864, the number of New Zealand plants established at Kew was fifty-three species of ferns and eighty-three seed plants, mainly trees and shrubs. John Smith, compiler of the list, remarked of the trees and shrubs that 'as they are not sufficiently hardy to withstand the open air of this climate, and, with few exceptions, their flowers being very inconspicuous, . . . they only rank as pretty evergreen greenhouse plants, chiefly to be found in Botanic Gardens only'.[56]

The number of New Zealand species of trees and shrubs in Kew Gardens had risen to about 140 by 1934, of which seventy kinds of *Hebe* and thirty species of other genera were regarded as sufficiently hardy to grow outdoors without any form of protection. Some examples of the hardy genera were *Carmichaelia* and *Notospartium* (native brooms), *Cassinia, Cyathodes, Hymenanthera, Leptospermum, Muehlenbeckia, Myrsine, Olearia, Senecio* and *Sophora*.[60]

THE FERN CRAZE

When John Smith began as a gardener at Kew in 1823, he specialized in ferns, raising many new kinds from spores obtained from dried herbarium specimens. Loddiges also had a considerable fern collection for sale, with many raised from spores, and the firm held a near monopoly in the fern craze that swept Victorian England. The fashion was made possible by the development of elegant models of the Wardian case, in which ferns could grow despite the smoke and fumes of the industrial towns.[35]

Dr J.H. Harvey of Frome, Somerset, who is an authority on plant nursery catalogues of the eighteenth and earlier centuries, kindly searched British lists to the end of the 1820s for New Zealand ferns, without success. Apparently New Zealand ferns were not offered for sale until the 1830s or later; however, the archaic nomenclature is difficult and the plant lists have many sections (stove, indoor, greenhouse, conservatory, etc.).[61]

Ferns were grown on a large scale by Mr Henderson of Towerville, near Glasgow, who used his vessels to bring plants from New Zealand; for example, in January 1866 when Captain Logan, master of the *Robert Henderson* was at Dunedin, he obtained tree ferns 'with stems 6, 8, and 10 feet in length and beautiful silvery fronds upwards to 6 feet in length' to take back to Scotland. Henderson grew tree ferns in pots outside in a sheltered corner of his garden until a new iron and glass fernery, heated in winter, was built to house them.[62]

The fern craze has ended, but the following New Zealand species remain popular: hen and chickens fern (*Asplenium bulbiferum*), shining spleenwort (*A. oblongifolium*), petako (*A. polyodon*), maidenhair (*Adiantum hispidulum* and allied species), kiokio (*Blechnum* species), Prince of Wales' feather (*Leptopteris superba*), heruheru (*L. hymenophylloides*), king fern (*Marattia salicina*), and crape fern (*Todea barbara*).

ALPINE PLANTS AND THE ROCKERY

Rockeries date from about 1774, but the first at Chelsea Physic Garden merely displayed geological specimens brought from Iceland by Sir Joseph Banks. Alpine plants for rockeries became popular about 1865 and as New Zealand alpines became available they were listed in British nursery catalogues. Kew botanists and artists described many of these novelties; for example, *Parahebe lyallii* was illustrated (as *Veronica lyallii*) in the *Botanical Magazine* for 1879, from a plant raised from seed by Mr Isaac Anderson-Henry of Edinburgh. At that time, the editor of the *Botanical Magazine* was Dr J.D. Hooker of Kew, who had named the species in 1854 in honour of Dr Lyall R.N. who had found it at Milford Sound, when *H.M.S. Acheron* was surveying the New Zealand coast. It was a most elegant addition to the alpine plants in cultivation in 1879.[63,64]

Many attempts were made to introduce *Celmisia* species into British gardens, but with little success; they grow better in Ireland although they are not easy to raise and are popular only with patient enthusiasts. Similar difficulties were experienced with *Ranunculus lyallii*, the well known Mount Cook buttercup. Many parcels of seed were sent to Britain before a plant was flowered at Kew in April 1886 and figured in the *Botanical Magazine* for the same year. In Ireland, it is a fine foliage plant, but seems to need very rich soil for flowering; in fact one gardener recommended planting it on a dung heap under a dripping tap.[65]

GREENHOUSE, CONSERVATORY, AND SUN LOUNGE
In the twentieth century, the development of the sun lounge from the Victorian and Edwardian conservatory provided a new home in Britain for New Zealand ferns, indoor plants and pot plants. Again a number of our species are lasting favorites; for example: rengarenga (*Arthropodium cirratum*) taupata or 'mirror plant' of some US lists (*Coprosma repens*), broad-leaved cabbage tree (*Cordyline indivisa*), ti pore — an introduced plant of the Māori found in Polynesia and Melanesia (*C. terminalis*), tree fern (*Cyathea dealbata* and allied species), whekī (*Dicksonia squarrosa*), puka (*Griselinia lucida*), N.Z. calceolaria (*Jovellana sinclairii*), mānuka (*Leptospermum scoparium* and dwarf cultivars), ramarama (*Lophomyrtus bullata*), kohukohu (*Pittosporum tenuifolium* and allied species and cultivars), houpara (*Pseudopanax lessonii*), nīkau palm (*Rhopalostylis sapida*), and patē (*Schefflera digitata*).

A surprise to New Zealanders is the inclusion of biddy-bid (*Acaena anserinifolia*, and allied species) in British and USA lists of indoor and ground-cover plants. Here we regard it as a pest, because the seeds catch in wool, but in Germany hanging baskets have been used to good effect as the flower heads, often red in colour, are ornamental.

The best of our climbers was introduced into Britain as *Clematis indivisa* about 1840, and the variety 'Lobata', now known as *C. paniculata* 'Lobata', has been a collector's item ever since. Outdoors in Britain it is destroyed in cold winters, and it must be grown under glass. A cultivar of it, named 'Stead's Variety', won a First Class Certificate when shown at the Royal Horticultural Society by Lord Aberconway on 1 May 1934 (Plate 1).[66]

Another surprise for New Zealanders visiting Britain is the use made of cabbage trees (*Cordyline* species), for tub plants both indoors and out, as well as for street planting and landscaping in the milder countries. In Volume 2 of the *Flora*, 1970, Moore and Edgar listed five cultivars of *Cordyline australis* which have been selected by European plant breeders, and probably more forms exist in northern hemisphere collections.[119]

Some New Zealand plants have proved suitable for peat gardening, both indoors and out; for example, *Bulbinella hookeri* and *B. rossii*, both sold as

houpara (*Pseudopanax lessonii*) **Araliaceae**
Engraved from a 1769-70 painting by Sydney Parkinson

'cats tails', *Cyathodes colensoi, Libertia peregrinans, Ourisia* species, and *Ranunculus lyallii. Drosera binata* and *D. spathulata*, sundews of Australia and New Zealand, are sometimes found in greenhouse collections of insectivorous plants.[67]

ENGLISH GARDENS OF NOTE
From time to time efforts have been made to list New Zealand plants in leading gardens; for example, Professor Arnold Wall published a 'preliminary catalogue' in the *Transactions of the Royal Society of New Zealand* for 1929. Among the outstanding gardens where he found collections of New Zealand plants were those of Hon. Vicary Gibbs, Aldenham House, Elstree, Herts; Major L. Johnson, Hidcote Manor, Gloucestershire; Mr Ingwersen, East Grinstead, Sussex; and Hon. Gerald Loder, Wakehurst Place, Ardingly, Sussex.[68]

In 1939, Dr H.H. Allan wrote an account of New Zealand plants in the garden of Dr A.H. Williams, near Horsham in Sussex. Dr Williams had about 200 species of New Zealand plants in cultivation, and Dr Allan included photographs of flowering specimens of *Notospartium carmichaeliae, Olearia avicenniaefolia, Myosotis spathulata*, and so on. Earlier, Dr Williams had written about his New Zealand plants in the *Journal of the Royal Horticultural Society* (1930), and *Gardening Illustrated* (1935). In the 1930 article, he wrote that repeated losses of some species were not due to the severity of frosts each year, but rather to the early resumptions of growth after winter and then late frosts which destroyed the new growth. In the 1935 paper he provided photographs of about twenty New Zealand plants, including novelties such as *Aciphylla squarrosa, Hibiscus diversifolius* and *Pachystegia insignis*.[69-71]

Graham Hutchins is mentioned later as a grower of *Hebe*. In 1980 he published a booklet on the New Zealand plants in cultivation at the County Park Nursery, Hornchurch, Essex. He explained how he had read *New Zealand Plants and their Story* by Leonard Cockayne, and then *New Zealand Alpine Plants* by Mark and Adams, and these so fired his enthusiasm that he flew to Auckland in January 1977 and spent six weeks collecting seeds and plants in the North and South Islands. He succeeded in bringing 294 species and cultivars into cultivation at Hornchurch, and his booklet lists them with notes on their hardiness, and so on. He expressed high hopes of introducing many of them into British gardens, and of obtaining further interesting plants from New Zealand.[72-74]

CORNWALL AND THE ISLES OF SCILLY

In 1896, F.H. Davey listed New Zealand plants that were acclimatized in the Falmouth-Truro district of Cornwall, including *Clematis indivisa* (now *C. paniculata*) *Clianthus puniceus, Corokia buddleioides, Edwardsia grandiflora* and *E. microphylla* (now *Sophora tetraptera* and *S. microphylla*), *Griselinia littoralis, Leptospermum scoparium, L. ericoides* (now *Kunzea ericoides*), *Metrosideros excelsa, Muehlenbeckia complexa, Myosotidium nobile* (now *M. hortensia*), *Olearia haastii, Phormium tenax, Pittosporum tenuifolium* and Veronicas (*Hebe* species).[75]

Captain Arthur Dorrien Smith joined an expedition to the Auckland and Campbell Islands in 1907 and visited the Chatham Islands in 1909 to gather plants and seed for the garden at Tresco Abbey, Isles of Scilly. From the Chatham Islands he obtained *Myosotidium hortensia* and several forms of *Olearia semidentata*, the most beautiful of the Olearias. By 1911, Dorrien Smith claimed to have 259 species of New Zealand plants 'thoroughly established' in the Tresco garden.

This famous collection was a source of seed and plants for Kew, of flowering specimens for the artists of the *Botanical Magazine*, and of horticultural news and photographs for the *Gardeners Chronicle* and similar journals. From these sources we learn that pōhutukawa (*Metrosideros excelsa*) formed part of the canopy of the garden, while karo and kōhūhū (*Pittosporum crassifolium* and *P. tenuifolium*) are used for hedges. These species are also used for hedges in Cornwall, from whence cut foliage of kōhūhū is sent to the London flower markets.[76]

HEBE IN ENGLAND AND SCOTLAND

From about 1840 to 1884, Isaac Anderson, later known as Anderson-Henry, an Edinburgh lawyer and amateur gardener, conducted pioneer experiments in grafting and hybridization. His aim apparently was to unravel the laws of genetics, and he corresponded with Charles Darwin. He succeeded in crossing two New Zealand species of *Hebe* with each other, and with another species from the Falklands. The New Zealand species were then known as *Veronica speciosa* and *V. salicifolia*. The Falkland Islands plant was known as *Veronica decussata*, now *Hebe elliptica*. He obtained several fine hybrids such as *Veronica 'Purple Queen'*, *V. lobelioides*, and *Hebe x andersonii*. By the 1890s, there were about forty New Zealand species of *Hebe* at the Royal Botanic Gardens, Edinburgh, as well as many hybrid forms.[77-79]

A recent British *Hebe* grower is Mr Graham Hutchins of the County Park Nursery at Hornchurch, Essex. He exhibited 120 species, varieties and hybrids of *Hebe* and *Parahebe* at the Royal Horticultural Society's Show at Westminster in 1976, and he was awarded the Lindley Silver-Gilt Medal. In 1979, he published a sixty page booklet, discussing topics such as the propagation, pruning, pests, and hardiness of each kind of *Hebe* in his collection. He recommended them for rock gardens, sink gardens, cavity walls, window boxes and tubs, and noted that, when they are planted 30-60 cm apart, they will completely smother all but the most vigorous of weeds. He recorded their medicinal virtues and added that they are much preferred by donkeys to oats.[80]

In February 1985, a Hebe Society was formed in Great Britain; its aims are to contact members and issue newsletters, to promote wider interest in the use of Hebes and allied plants, to improve and increase the plants available, and to control naming and assist identification. The Society is forming a National Reference Collection at Trewidden, Cornwall, with the aid of Lawrie Metcalf of Invercargill who has prepared a checklist of cultivars. At the beginning of 1987, the Society had 150 members in Great Britain, Ireland, USA, New Zealand etc.; it is the first British or international society devoted to a group of New Zealand plants.

In 1988, three publishers (Christopher Helm of London, Timber Press of Portland, and Caxton Press of Christchurch, New Zealand) published *Hebes and Parahebes* by Douglas Chalk, one of the founders and Chairman of the Hebe Society. The book includes chapters on growing, propagation, and choosing Hebes, brief diagnoses of the numerous species, varieties and cultivars, and an indication of the hardiness of each kind. The coloured and black and white illustrations are most attractive. Some of the cultivars were bred in New Zealand and others in France, Ireland, England, and Scotland. Leading British gardens and gardeners commemorated in *Hebe* cultivar names include 'Kewensis', 'Edinensis', 'Highdownensis', 'Warleyensis',

'Boscowensis', and 'Bowles Hybrid'. Appendix 2 records the results of the 1982 Royal Horticultural Society's trial of *Hebe* species and cultivars at Wisley, and Appendix 3 lists some British and New Zealand gardens with noteworthy *Hebe* collections.[81]

3.8 Scotland

Mention has been made of Mr Isaac Anderson-Henry of Edinburgh, who conducted pioneer experiments in crossing New Zealand species of *Hebe*; of the great collections of *Hebe* species and cultivars at the Royal Botanic Gardens, Edinburgh; and of Mr Henderson of Towerville near Glasgow, who grew tree ferns on a grand scale.

Dr W. Lauder Lindsay, an Edinburgh medical man and botanist, visited New Zealand in the summer of 1861-2 and spent about four months with William Martin, nurseryman, of Fairfield near Dunedin. In 1862, Lindsay reported to the British Association on the heavy losses experienced by New Zealand settlers when their stock ate tutu (*Coriaria*). His main paper on the taxonomy of higher plants of Otago was published in 1868, with coloured illustrations by the famous botanical artist, W. Fitch. In June 1878, in a paper to the Botanical Society of Edinburgh, he discussed the growth of New Zealand plants in Great Britain, especially the results obtained with Otago plants in various gardens of Scotland.[40,62,82,83]

Another northern plantsman of note was Mr William Gorrie of Rait Lodge, Trinity, who was a leading grower of New Zealand plants. He believed that hardy plants suitable for conditions in Scotland would be found in the South Island mountains and, as early as 1863, he tried to persuade other gardeners to join in sending a collector there. For many years he imported seed of New Zealand plants from Canterbury and Otago, and asked his suppliers to collect from the colder and more exposed habitats. In 1880, as President of the Edinburgh Botanical Society, he gave a paper on the hardiness of New Zealand plants, listing thirty which had withstood the unusually severe winters of 1878-79 and 1879-80. He remarked that tough branches of ribbonwood (*Plagianthus betulinus*, now *P. regius*) made serviceable garden twine, the wood showed promise for paper-making and fashioning barrel hoops and basket handles, while small twigs made excellent points for fishing rods. He considered that the tough leaves of *Libertia* should be of interest to paper-makers, that the hard, beautifully mottled and veined timbers of large growing Olearias were useful for inlaying and veneering, and the wind resisting cabbage tree (*Cordyline australis*) made an excellent summer pot-plant — in winter it should be kept under glass without heat.[84,85]

The value of Scottish collections of New Zealand plants was demonstrated in 1982, when the late Mr A.W. Purdie, Botany Division, DSIR, Lincoln, visited Great Britain to study historical collections of native brooms; in the rockery of the Royal Botanic Gardens, Edinburgh, he found a plant of *Carmichaelia prona*. This species once grew at Lake Lyndon and Lake Merrymere and at the mouth of the Rakaia River, but is now thought to be extinct in New Zealand.[86]

3.9 Ireland

New Zealand plants which barely survived the rigours of winter in England and Scotland fared much better in milder and wetter southern Ireland. When the Botanic Gardens at Glasnevin, Dublin, first opened to the public in 1800, the plant collection included New Zealand flax (*Phormium tenax*), kōwhai (*Sophora tetraptera*), and mānuka (*Leptospermum scoparium*). In Irish gardens today, flax serves as an effective windbreak on the windswept west coast, kōwhai is rare, and mānuka is ubiquitous in sheltered places.[87,90]

In the early years of last century, losses of New Zealand plants were heavy during severe winters; however, *Cordyline australis* proved hardy, several *Pittosporum* species were 'tolerant' and *Griselinia littoralis* was proving something of a weed in sheltered places. Rodger, McClelland & Co., of Newry, County Down, was probably the first Irish nursery to offer a number of New Zealand plants for sale (including eighteen species and cultivars of *Hebe* in 1879 and 1880).[90]

From 1870 to 1911, William Gumbleton at the Belgrove Estate on an island in Cork Harbour, and, later, Lord Talbot de Malahide at Malahide Castle, collected New Zealand tree daisies (*Olearia* species), and their success led other Irish gardeners to appreciate the value of these plants. Most of the larger gardens had, and still have, a quota of *Olearia*, with cabbage trees (*Cordyline*), and toetoe (*Cortaderia*).

Accounts of gardens of north-west Ireland mention New Zealand lacebark (*Hoheria populnea*) seeding freely, and kōhūhū (*Pittosporum tenuifolium*) and tarata (*P. eugenioides*) as common. Tree-ferns grow well, and at Rossdohan, County Kerry, ponga (*Cyathea dealbata*) reproduces freely. In Dublin and other towns, hedges of broadleaf (*Griselinia littoralis*) have been described as far too common.[88,89]

Many attempts were made last century to grow New Zealand flax (*Phormium tenax*) as a commercial crop in Ireland, but no industry resulted. However, flax has been used very effectively in the south and west as a windbreak around gardens, to provide shelter for tender exotics. Other New Zealand species used for the same purpose are *Griselinia littoralis, Olearia macrodonta, O. traversii*, and *Hebe* cultivars.[90]

In April and May 1988, Dr E. Charles Nelson, National Botanic Gardens, Glasnevin, Dublin, visited New Zealand. In an address to the Institute of

Horticulture, Christchurch, he spoke of New Zealand's contribution to Ireland's garden heritage. He listed more than 200 species of trees and shrubs native to New Zealand and cultivated in Irish 'heritage gardens'. The list includes plants which are regarded in New Zealand as 'endangered', for example, *Carmichaelia williamsii, Clianthus puniceus, Hebe cupressoides, H. speciosa, Olearia pachyphylla, Pittosporum dallii,* etc. The list does not include some threatened New Zealand herbaceous plants that are common in Irish gardens. It seems that the time is near when we will look in Ireland for threatened and extinct New Zealand plants.[95]

During his visit, Nelson spoke of Henry Hammersley Travers (1844-1928), a lawyer of Irish origin who, in his retirement, greatly enriched the plant collections of the Botanic Gardens at Glasnevin. Travers explored the Chatham Islands in 1863 and 1871, and recorded customs of the Chatham Islanders and the devastation caused to the indigenous flora by fires and introduced animals. In the early 1900s Travers collected seeds and live plants in the Wellington, Nelson and Canterbury ranges. Nelson gave a sympathetic summary of his letters to Glasnevin and listed the many seeds and plants he sent between about 1906 and 1914. Today, thanks largely to Travers, the following Chatham Island plants (several of which are now rare and threatened) are well represented in the National Botanic Gardens at Glasnevin and in other Irish gardens:

Astelia chathamica, Corokia macrocarpa, Geranium traversii, Myosotidium hortensia, Olearia chathamica, O. traversii, the form of *O. semidentata* now known as *Olearia* 'Henry Travers', and *Senecio huntii.*[91,95]

Nelson also listed more than twenty-five cultivars that have been selected or bred in Ireland from endemic New Zealand plants. Examples are:

1. *Pittosporum tenuifolium* 'Silver Queen', one of the loveliest cultivars of that ornamental shrub, was noticed as a bud sport and propagated by James Coey, who established the Slieve Donard Nursery at Newcastle in 1910. It was given an Award of Merit by the Royal Horticultural Society of London in 1914. There is an exquisite painting of it by Wendy Walsh, with text by Charles Nelson, in *An Irish Florilegium* (1983).[92,94]

2. *Olearia* 'Henry Travers' was sent from Wellington to the Botanic Gardens, Glasnevin, by Henry Travers in 1908, and flowered in 1910. It differs from the species *O. semidentata,* with which it was identified for many years, and is thought to be of hybrid origin.[93]

3. *Hebe* 'Headfortii' was raised by the Marquis of Headfort in his garden at Kells, County Meath, and distributed about 1930. The sender of the seed to Kells is not known. There is nothing in the wild in New Zealand quite like it, and its relationships are not clear. However, it is a very fine garden plant.[93,94]

Paintings by Wendy Walsh of these two cultivars, with text by Charles Nelson, were published in *An Irish Florilegium* 2 (1987).

3.10 France

A. Thouin, First Gardener at the Jardin de Plantes, provided directions for the gardener attached to the voyage of the Comte de la Perouse, who explored the Pacific from 1785 until 1788. Thouin also established an 'école des plantes d'usage dans l'économie rurale et domestique'. In 1803, he wrote a paper on *Phormium tenax* in which he recorded that 'M. le Chevalier Banks', on his return from the South Seas, sent a quantity of seed to the garden, but none germinated. Then Mr W.T. Aiton of Kew forwarded a plant in 1800, and Grimwood, 'fleuriste à Kinsington' sent another in 1802.[96]

In 1803, the corvette *Le Naturaliste* returned to France with many 'pieds' (stolons or fans), obtained from the Governor's garden at Port Jackson; these had come originally from Norfolk Island. In 1812, Faujas de Saint Fond reported the first flowering of *Phormium tenax* in May, at the garden of M. Freycinet in the department of La Drôme. No fruits set, but the artist, M. Turpin, included Gaertner's figures of *Phormium* fruit in a lovely illustration accompanying Saint Frond's paper in *Annales — Muséum National d'Histoire Naturelle* for 1812.[97]

Phormium tenax flowered and set fruit for the first time in France at the Paris garden of M. Boursaut, and Pierre-Joseph Redouté made a superb etching for the eight volume *Les Liliacées*, issued between 1802 and 1816 under the patronage of the Empress Josephine. The plate of New Zealand flax is dated 1815.[98]

Between 1803 and 1815, more than 200 species of exotic plants were flowered at Malmaison for the first time in France, and many for the first time in Europe. Josephine employed Redouté to illustrate the flowers in immense folio works, which are a magnificent monument to her enthusiasm and extravagance, and the despair of librarians who have their care (the books are too large, and some are bibliographic nightmares). The only New Zealand plant we have traced to Malmaison is New Zealand spinach (*Tetragonia tetragonioides*), taken back to England by Banks in 1771, and etched by Redouté in 1803.[99] *Apium prostratum* was etched in 1804, but it was raised from Tasmanian seed.

In August 1839, Captain Doucet, commander of a Havre whaler, collected another species of *Phormium* 'de la baie Chaldy, non loin de la baie de Préservation'. The plants flowered and set seed at Chile on the way home, and again at Cherbourg in 1847. Auguste Le Jolis, Secretary of the Cherbourg Horticultural Society, described and illustrated the new species in *Revue Horticole* for 1848, and named it *Phormium cookianum* in a long paper on the different kinds of *Phormium*, and published in the *Bulletin de la Société d'Horticulture de Cherbourg* for the same year (Plate 4).[100,101]

J.S.C. Dumont d'Urville, who visited New Zealand in 1824, 1826-1827, and

New Zealand spinach (*Tetragonia tetragonioides*) **Aizoaceae**
Engraving: P.-J. Redouté (1803)

1840, took or sent back seeds to the Natural History Museum, Paris, and probably E.F.L. Raoul, who was stationed at Akaroa as surgeon to the French corvettes *L'Aube* and *L'Allier* from 1840 to 1843, also took or sent seeds home.

P.J. St. C.C. de Belligny, Doctor of Natural History, and travelling correspondent of the King's Botanical Gardens, who reached Akaroa on the *Comte de Paris* in August 1840, also sent seeds and live plants back to France and arranged for the seeds 'to be germinated by ingenious means' during the voyage. One container of plants from Akaroa Bay was forwarded to A. Brongniart, Professor of Natural History at the Museum of Natural History, Paris, on 18 February 1843, but whether the contents were live or dried is not clear. M. Simpson, whose researches on Belligny were published in 1984, commented 'it is tantalising to think that information regarding these and other collections may be stored away somewhere in France'.[102]

As early as 1872, Dr J.D. Hooker of Kew advised Thomas Kirk, who was revising the New Zealand species of *Pittosporum*, that several of them under cultivation in the south of France had produced self-sown hybrids. However, except in favoured localities, species of *Pittosporum* such as *P. tenuifolium* and *P. eugenioides* are treated as greenhouse ornamentals.[103]

Other New Zealand plants grown as greenhouse ornamentals in France include flax (*Phormium tenax*), but it is naturalized on the Brittany Coast, cabbage trees (*Cordyline australis* and *C. indivisa*), nīkau palm (*Rhopalostylis sapida*), Kirk's daisy (*Brachyglottis kirkii*) mānuka (*Leptospermum scoparium*) and some Hebes.

In 1986, Luc Decourtye, head of a laboratory for research on ornamental plants at Angers in the Loire Valley, spent five months with Botany Division at Lincoln near Christchurch. On the 21 April he gave a seminar on ornamental shrub breeding. While in New Zealand, he had collected seeds, cuttings, and plants of alpine and subalpine species to evaluate their cold hardiness under French conditions. The aim is to find alternatives for widely-used ornamentals which have been affected by fireblight and other diseases in France. Loss of plants can spoil a hedge, so hedge plants must be reliable. He hoped that genera such as *Leptospermum*, *Pittosporum*, *Olearia*, and *Coprosma* would provide suitable hedge plants, but they must be bred to survive frost.[104]

Leptospermum should be a popular outdoor plant in France; however, the plants tried there came mainly from the warmer North Island and died during severe French winters. Apparently none of the South Island high country forms have been tried.

In 1989, Dr Warwick Harris, then Director of Botany Division, spent six months at Angers to study the New Zealand plants that had experienced two French winters. As well as the Angers collection, Luc Decourtye had trial plantings at Frejus on the Mediterranean, and on the Brittany Coast. One interesting observation Dr Harris made was that plants on the Brittany Coast, in a milder climate, suffered more severe damage in winter than those at Angers. This suggested that the extra 'ripening' in the hotter Angers summer somehow conditioned the plants to tolerate colder winters.

Harris listed other New Zealand genera being tried at Angers as New Zealand brooms (*Carmichaelia*, and allied genera), mountain forms of tauhinu (*Cassinia*), lacebarks (*Hoheria*), Marlborough daisies (*Pachystegia*) from the Botany Division collection, lancewoods (*Pseudopanax*), southern beeches (*Nothofagus*), kōwhai, (*Sophora*), particularly the prostrate forms studied at Botany Division, *Coprosma* and *Corokia* for coloured berries, and *Hebe* for form, leaf variety, and coloured flowers.[105,106]

3.11 USA

In March 1840, William Dunlop Brackenridge (1810-1893) visited New Zealand as Assistant Botanist and Horticulturist to the US Exploring Expedition commanded by Charles Wilkes of the United States Navy. The expedition lasted from 1838 to 1842, and Brackenridge collected 10,000 dried plants as well as seeds and live plants thoughout the Pacific. The dried plants formed the first collection of the US National Herbarium, and the conservatory of live plants which Brackenridge maintained ultimately led to the formation of the USDA Plant Introduction Service.[107,108]

New Zealand plants were introduced into Californian gardens from the middle of last century. William C. Walker opened the Golden Gate nursery in San Francisco in 1849, and offered for sale exotic plants from the South Pacific and Australia. He bought seed from Guilfoyle's Exotic Nursery at Double Bay, Sydney, and exhibited *Clianthus puniceus* at a horticultural show in 1854. By 1874, Bernard S. Fox of San Jose offered *Areca sapida* (now *Rhopalostylis sapida*), F. Ludermann's Pacific Nurseries of San Francisco sold *Cordyline indivisa* and *C. forsteri* (probably *C. australis*), and James Hutchinson's Bay Nursery at Oakland listed *Corynocarpus laevigatus*, *Hebe speciosa*, and *Pittosporum tenuifolium*.[109]

In 1939, W.E. Lammerts, University of California, Los Angeles, fertilized flowers of *Leptospermum scoparium* 'Nichollsii' with pollen from the 'Rose Double' cultivar. Very dwarf, purple-stemmed, purple-leaved forms, almost all with double flowers, resulted from the cross, and Lammerts bred another generation from them. He raised 850 seedlings, the variability of which was greater than that of the original parents. Several horticulturally interesting forms were then chosen, named, and released to nurserymen. It is to the shame of New Zealand that no similar work has been attempted here.[110,111]

In present day Californian gardens, ground cover plants used may include *Cotula squalida*, sold as 'N.Z. Brass buttons', *Muehlenbeckia axillaris*, sold as 'M. nana' or 'Creeping wire vine', and *Dichondra micrantha* from Cuba, sold as 'D. repens' or 'Mercury Bay weed'.[112,113]

3.12 Australia

Mr Charles Fraser, a private in the 46th Regiment, was appointed 'Superintendent of the Botanic Gardens' in 1816, and 'Colonial botanist' a year later, the Governor's Garden at Sydney having been renamed the Botanic Gardens about this time. In 1826 he visited New Zealand in *H.M.S. Larne*, from 19 to 28 February and took back 'many living examples of the vegetable products to establish in the Botanic Garden at Sydney'.[114]

In the lists of plants cultivated in the Gardens during Fraser's time, there are a number of New Zealand plants; for example, *Cordyline australis* 1822, *Coriaria sarmentosa* 1819, *Edwardsia microphylla* 1820, *Metrosideros species* 1819, *Phormium tenax* 1820, etc. Fraser is credited with New Zealand plants in 1819, 1822, 1823, 1825, etc., but it is unlikely that he visited this country in those years.

Fraser's reports for 1827 and 1828 mention several New Zealand plants. In the list of esculent vegetables and pot herbs cultivated in December 1827 was *Tetragonia expansa*, N.Z. spinach. Fruits which have come into bearing during the year 1828 included *Corynocarpus laevigatus* from New Zealand. Exotic forest trees cultivated in the Garden in December 1828 included *Podocarpus* sp., 'Cowee'; and *Agathis* sp., 'Cowee-cowee'. Plants cultivated in the garden which are used for commerce in 1828 included *Phormium tenax*, N.Z. flax, and a red var., Dr Fairfowl, 1820.

Both Allan Cunningham and his brother, Richard Cunningham collected live plants for the Botanic Gardens, Sydney, and Allan arranged shipments to and from the Royal Gardens, Kew. Allan visited New Zealand in 1826-1827 and in 1838, and Richard Cunningham visited in 1834; both took live material back to Sydney.

In the accessions for later years, donors to the Sydney Botanic Gardens included: 1849, Mr Swainson, case of plants, with list of desiderata; 1853 J. Ludlum Esq., case of seedling plants and case of ferns; 1878, Dr J. Hector, seeds of *Areca sapida* (now *Rhopalostylis sapida*), *Elaeocarpus dentatus, Parsonsia albiflora* (now *P. heterophylla*), and *Pittosporum eugenioides*.

In 1845-6, the Sydney Gardens sent live plants of *Dacrydium cupressinum* to the Botanic Gardens, Calcutta, to the Professor of Botany at Paris, and to Cape Town, Cape of Good Hope; *Phormium tenax* to Samoa and Tahiti; and *Corynocarpus laevigatus* to Cape of Good Hope, Hobart Town, Melbourne and Hong Kong. Later there were active botanic gardens at Brisbane, Melbourne, Adelaide, and Hobart, as well as Sydney, and it seems that these institutions distributed some of the common New Zealand plants in their exchanges.[59]

John Carne Bidwill (1815-1853), an Australian settler, visited New Zealand in 1839 and was the first European to climb Ngauruhoe. He returned to Port Nicholson in 1840, and published an account of his visits to New Zealand in the following year. During 1847, he served for a few months as Government Botanist at Sydney, and in 1848 he collected plants in the Nelson mountains. He established some New Zealand plants in the Sydney Botanic Gardens, and corresponded with the Hookers of Kew who named the new alpine plants that he discovered in New Zealand.[115]

From 1871 to 1878, Ferdinand von Mueller of the Botanic Gardens, Melbourne, prepared five papers for the Victorian Acclimatisation Society on 'industrial plants available for culture in extra-tropical countries', and these lists were republished in his classic *Select Extra-tropical Plants*, which went through many editions. The Baron listed more than forty useful plants from New Zealand, but none of them has served as the basis for a new industry in Australia.[116]

Generally, New Zealand species are outclassed in the garden by more striking Australian plants, once described as a flora in fancy dress, but M. Bulfin, Botany Division, DSIR, Lincoln, commented in 1978, following a visit to Australia, that 'gardeners in both Tasmania and Melbourne were proud to show New Zealand plants they grow well. In particular, Raoulias are prized as ground cover and some beautiful *Celmisia* plants were evident'.[117]

1. Beaglehole, J.C. (ed.) *The Journals of Captain James Cook*, Cambridge University Press for the Hakluyt Society. Vol. 1. *The Voyage of the Endeavour 1768-1771* 1955. 684p. 2nd ed. 1968.
2. Beaglehole, J.C. (ed.) *The Endeavour Journal of Joseph Banks 1768-1771.* Sydney (Trustees of the Public Library of N.S.W. & Angus & Robertson) 1962, 2 vols.
3. Nelson, E.C. (1983) 2(4) *Telopea* 347.
4. Stearn, W.T. *A History of the British Museum (Natural History) 1753-1980.* London (Heinemann) 1981, 414p.
5. Aiton, W. *Hortus Kewensis, or a Catalogue of Plants Cultivated in the Royal Botanic Garden at Kew.* London (George Nicol) 1789, 3 vols.
6. Aiton, W.T. *Hortus Kewensis* London (Longman) 1810-1813. 2nd enl. ed. 5 vols.
7. Miller, J. (Mueller, J.S.) *Icones novae*, Londini (John Miller) 1780, 7 pl. 1. *Sophora tetraptera*, 2. *Phormium tenax.*
8. Curtis, W. (1791) 5 *Bot. Mag.* 167. *Sophora tetraptera.*
9. Ferguson, R. (1990) 1 *Horticulture in New Zealand* 24.
10. Beaglehole, J.C. (ed.) *The Journals of Captain James Cook* Cambridge University Press, Vol. 2. *The Voyage of the Resolution and Adventure 1772-1775* 1961. 1021p.
11. Hoare, M.E. *Three Men in a Boat. The Forsters and New Zealand Science* Melbourne (Hawthorn Press) 1975, 16 p.
12. Begg, A.C., Begg, N.C. *Dusky Bay* Christchurch (Whitcombe & Tombs) 1975, 3rd ed. 240p.
13. Dawson, W.R. *The Banks Letters* London (British Museum — Natural History) 1958, 905 p.
14. Britten, J. (1916) 54 *J. Bot.* 345. (1917) 55 ibid. 54.
15. Murray, J.A. (1781) 3 *Commentat. Soc. Regiae Sci.* Gottingae 1780 3, t. 1 *Cercodia erecta.*
16. Diment, J.A., Wheeler, A. (1984) 11(3) *Archives Nat. Hist.* 457. Solander, D.C. 'Seeds from the South Seas 1780', p. 469.
17. Forster, J.R., Forster, G. *Characteres Generum Plantarum* London (White, Cadell & Elmsley) 1775, 1776. 1st ed. 75 p., 2nd ed. 150 p., 3rd ed. 76 p.
18. Solander, D.C. [1769-1770] 'Primitae Florae Novae Zelandiae' Original in the Botanical Manuscripts of the Natural History Museum, London; typescript copy in the Auckland Institute and Museum Library.
21. Pallas, P.S. *Enumeratio Plantarum quae in Horto viri illustris atque excell. Dni. Procopii A Demidof* Petropoli (Acad. Imder. Scientiarum) 1781. p. 150, t. 1 *Demidovia tetragonoides.*
23. Murray, J.A. (1785) 6 *Commentat. Soc. Regiae Sci.* Gottingae 1783 *Commentat. Phys.* 3. p. 13, t. 5 *Tetragonia expansa.*
25. Schrader, H.A., Wendland, J.C. *Sertum Hannoveranum seu Plantae Rariores* Gottingae (Vandenhoeck et Ruprecht) 1796, vol. 1, fasc 2, p. 25, t. 15 — *Melaleuca scoparia* var. *diosmatifolia* and var. *myrtifolia.*
26. Roemer, J.J., Usteri, P. (1790) 3(7) *Magazin für die Botanik*, Zürich. 177, t. 2. *Philadelphus floribundus.*
27. Scopoli, J.A. *Deliciae florae et faunae Insubricae* Paviae (Monasterii S. Salvatius) 1786-1788. 3 vols. Vol.1, p. 32, t. 14 *Tetragonia expansa.*
30. Jacquin, N.J. *Plantarum rariorum horti caesarei Schoenbrunnensis Descriptiones et Icones* Wiennae (C.F. Wappler) 1797-1804. 4 vols. Vol. 3, p. 17, t. 269 *Sophora microphylla* (1798).

34. McNab, R. *Murihiku and the Southern Islands* Invercargill (W. Smith) 1907. 377 p. 'Menzies' Journal' p. 301.
35. Allen, D.E. *The Victorian Fern Craze: A History of Pteridomania* London (Hutchinson) 1969, 83p.
36. Shepherd, W, Cook, W. *The Botanic Garden Wellington. A New Zealand History 1840-1987* Wellington (Millwood Press) 1989, 396 p.
37. Brown, (1850) Letter to Chief Justice [W. Martin], 1 April 1850, in the Brown & Campbell letterbooks, in the Auckland Institute and Museum Library.
38. Challenger, S. (1978) 6 *Annual Journal Royal N.Z. Inst. Hort.* 139.
39. Newman, J. (1853) Letters to W. Newman, Louth, in Auckland Institute and Museum Library.
40. Cheeseman, T.F. *Manual of the New Zealand Flora* Wellington (Government Printer) 1906 ed. 1, 1199p. 1925 ed. 2, revised by author, edited by W.R.B. Oliver 1163p.
41. Forster, G. *De Plantis Esculentis* Berlin (Haude & Spener) 1786a 80p.
42. Forster, G. *Florulae Insularum Australium Prodromus* Gottingae (J.C. Dieterich) 1786b 103p.
43. Hemsley, W.B. (1890) 3, 7 *Gardeners Chronicle* 255. Sat. 1 March.
44. Gaetner, J. *De Fructibus et Seminibus Plantarum* Stuttgart (Academie Caroline) 1788-1791, 2 vols.
46. Galpine, J.K. *The Georgian Garden. An Eighteenth-Century Nurseryman's Catalogue* Standbridge (Dovecote Press) 1983. 56p. Introduction and commentary by John H. Harvey.
47. Harvey, J.H. *Early Nurserymen* Chichester (Phillimore) 1974. 276p.
48. Harvey, J.H. personal commun. 13 July and 7 August 1984.
49. Curtis, W. (1796) 10 *Bot. Mag.* 349 *Solanum laciniatum*.
50. Andrews, H.C. (1812) 10 *Botanist's Repository* 622. *Leptospermum scoparium*.
51. Sims, J. (1822) 50 *Curtis's Bot. Mag.* 2362 *Tetragonia expansa*.
52. Lindley, J. (1835) 1 n.s. *Trans. Hort. Soc.* 519. t. 22 *Clianthus puniceus*.
53. Lindley, J. (1836) 8 n.s. *Edwards's Botanical Register* 1775 *Clianthus puniceus*.
54. Hooker, J.D. (1886) 112 *Curtis's Bot. Mag.* 6864 *Colensoa physaloides*.
55. Sampson, F.B. *Early New Zealand Botanical Art* Auckland (Reed Methuen) 1985. 142p.
56. Smith, J. (1880) *Records of the Royal Botanic Gardens, Kew* 71. Record II The New Zealand Collection.
57. Cooper, R.C. (1970) 7 *Rec. Auckland Inst. Mus.* 123.
58. Hooker, W.J. (1843) 70 *Curtis's Bot. Mag.* 4057 *Veronica speciosa*.
59. Royal Botanic Gardens, Sydney. (1828-1847) Vol. B1 Plants and Seeds received; (1832-1847) Vol. B2 Plants and Seeds distributed; (1852-1869) Vol. B3 Plants and Seeds distributed; (1852-1870) Vol. B4 Plants and Seeds received; (1870-1880) Vol. B5 Plants and Seeds received.
60. Royal Botanic Gardens, Kew. *Hand-list of Trees and Shrubs (including Coniferae) cultivated in the Royal Botanic Gardens, Kew.* London (H.M. Stationery Office) 1934. 4th ed. 362 p.
61. Sweet, R. *Hortus Britannicus or, A Catalogue of Plants, Indigenous, or Cultivated in the Gardens of Great Britain* London (James Ridgway) 1827 1st ed., 1830 2nd ed. 2 vols.
62. Lindsay, W.L. (1878) July 20 *The Gardeners' Chronicle* 79.
63. Gorer, R., Harvey, J.H. (1979) 7 (2) *Garden History* 69.
64. Hooker, J.D. (1879) 105 *Curtis's Bot. Mag.* 6456. *Veronica lyallii*.

65. Hooker, J.D. (1886) 112 *Curtis's Bot. Mag.* 6888 *Ranunculus lyallii.*
66. Lloyd, C. *Clematis* London (Collins) 1977 208p.
67. Evans, A. *The Peat Garden and its Plants* London (Dent & Sons, with the Roy. Hort. Soc., London) 1974 164p.
68. Wall, A. (1929) 60 *Trans. Roy. Soc. N.Z.* 379.
69. Allan, H.H. (1939) 9(2) *J. Roy. N.Z. Inst. Hort.* 29.
70. Williams, A.H. (1930) *J. Roy. Hort. Soc.* 101.
71. Williams, A.H. *New Zealand and her Plants: Random Reveries from a Sussex Garden* Reprint of articles from *Gardening Illustrated* [1935?] 24p.
72. Hutchins, G. *New Zealand Plants. A Descriptive List of Species and Cultivars as in Cultivation at County Park Nursery, Hornchurch, Essex.* 1980. 58p.
73. Cockayne, L. *New Zealand Plants and their Story* Wellington (Government Printer) 1927. 3rd ed. 271p.
74. Mark, A.F., Adams, N.M. *New Zealand Alpine Plants* Wellington (Reed) 1979. 2nd ed. 265p.
75. Davey, F.H. (1896) 44 *J. Roy. Inst. Cornwall* 1.
76. King, R. Tresco, *England's Island of Flowers* London (Constable) 1985. 160 p.
77. Anon. (1884) September 27 *The Gardeners' Chronicle* 400 Obituary — Isaac Anderson-Henry.
78. Lindsay, R. (1894) June 16 *The Garden* Pl. 966 *Veronica* 'Purple Queen'.
79. Souster, J. (1956) 81 *J. Roy. Hort. Soc.* 495. *Veronica* 'Blue Gem'.
80. Hutchins, G. *Hebe and Parahebe. A Descriptive List of Species and Cultivars as in Cultivation at County Park Nursery, Hornchurch, Essex.* 1979, 60p .
81. Chalk, D. *Hebes and Parahebes* Christchurch (Caxton Press) 1988, 152p.
82. Lindsay, W.L. (1862) *British Association Proceedings Annual Report* 98.
83. Lindsay, W.L. *Contributions to New Zealand Botany* London (Williams & Norgate) 1868.
84. Gorrie, W. (1880) 14 *Trans. Edinburgh Bot. Soc.* 52, 106.
85. Gorrie, W. (1880) January 17 *The Gardeners' Chronicle* 75, February 7, 171.
86. Purdie, A. (1982) 81 *Botany Division Newsletter* 5.
87. Nelson, E.C., McCracken, E.M. *The Brightest Jewel. A History of the National Botanic Gardens, Glasnevin, Dublin* Kilkenny (Boethius Press) 1987, 275p.
88. Morley, B.D. 'The Contribution of Southern Hemisphere Plants to Irish Gardens', p. 153 In Brady, A., Nelson, E.C. (eds) *Irish Gardening and Horticulture* Dublin (Royal Hort. Soc. Ireland) 1979.
89. Morley, B.D., Nelson, E.C. (1979) 7(3) *Garden History* 53.
90. Nelson, E.C. (1989) 16 *Annual Journal Royal N.Z. Inst. Hort.* 4.
91. Nelson, E.C. (1989) 16 ibid. 60.
92. Walsh, W., Ross, R.I., Nelson, E.C. *An Irish Florilegium* London 1983.
93. Walsh, W., Nelson, E.C. *An Irish Florilegium II. Wild and Garden Plants of Ireland* London (Thames & Hudson) 1987, 216p.
94. Nelson, E.C., Walsh, W. *An Irish Flower Garden* Kilkenny (Boethius Press) 1984. Text by E.C. Nelson, illustrations by W. Walsh. 218 p.
95. Wilson, C.M., Given, D.R. *Threatened Plants of New Zealand* Wellington (DSIR Publishing) 1989. DSIR Field Guide. 151 p.
96. Thouin, A. (1803) 2 *Annales du Muséum National d'Histoire Naturelle* 228.
97. Saint Fond, Faujas de (1812) 19 *Annales du Muséum National d'Histoire Naturelle* 419, pl. 20 *Phormium tenax.*
98. Redouté, P.-J. *Les Liliacées* Paris 1802-1808 [-1816]. 8 vols, 486pl. 1815 Pl. 448-9 *Phormium tenax.*

99. Candolle, A.P. de *Plantarum historia succulentarum. Histoire des plantes grasses . . . dessinées par P.-J. Redouté* Paris 1798-1837. 32 parts, 182 pl. pl. 114 *Tetragonia expansa* (1803).

100. Le Jolis, A. (1848) 2 *Bull. Soc. Horticulture de Cherbourg* 53.

101. Le Jolis, A. (1848) Ser. 3, 2 *Revue Horticole* 4.

102. Simpson, M.J.A. 'The Contribution of the French to the Botany of Banks Peninsula', p. 65 in Hoare, M.E., Bell, I.G. (eds) *In Search of New Zealand's Scientific Heritage* Wellington (The Royal Society of New Zealand) 1984. Bull. 21. 123 p.

103. Kirk, T. (1872) 4 *Trans. N.Z. Inst.* 267.

104. Decourtye, L. (1986) 108 Botany Division Newsletter 20.

105. Harris, W. (1989) 129 *Botany Division Newsletter* 3.

106. *New Zealand Herald*, 30 March 1989 — Visit of W. Harris to France.

107. Harrison, J. (1843) *The Floricultural Cabinet and Florists' Magazine* 243.

108. Wilkes, C. *Narrative of the United States Exploring Expedition during the years 1838, 1839, 1840, 1841, 1842* London (Whittaker) [1845]. 372p.

109. Streatfield, D.C. (1984) 12(1) *Garden History* 58.

110. Lammerts, W.E. (1945) 6 *J. Calif. Hort. Soc.* 250.

111. Lammerts, W.E. (1946) September *Pacific Coast Nurseryman* 10.

112. Forde, M.B. (1978) 16 *N.Z.J. Bot.* 283.

113. Sunset Editorial Staff *Lawns and Ground Covers* Menlo Park, California (Lane Books) 1970. 112p.

114. Maiden, J.H. (1908) 42 *J. Proc. Roy. Soc. N.S.W.* 60, p. 103 — C. Fraser.

115. Bidwill, J.C. *Rambles in New Zealand* London (W.S. Orr & Co) 1841. 104 p. Capper Press reprint 1974.

116. Mueller, F. von *Select Extra-tropical Plants* Sydney (Government Printer) 1881. 403 p. (And many other editions).

117. Bulfin, M. (1978) 39 *Botany Division Newsletter* 7.

118. Sexton, R. *H.M.S. Buffalo* Adelaide (Australasian Maritime Historical Society) 1984. 183p.

119. Moore, L.B., Edgar, E. *Flora of New Zealand Volume 2* Wellington (Government Printer) 1970. 354p.

4

Native Plants in New Zealand Gardens

4.1 Māori Gardening

Helen Leach (1984) praised the gardening skill of the early Polynesian arrivals in New Zealand, the ancestors of the Māori people, who brought kūmara, taro, yams, gourds, paper mulberry, and ti pore (*Cordyline terminalis*) from their tropical homeland, and kept them in cultivation at favourable sites in northern New Zealand for perhaps 1000 years. Where necessary, they built fences and stone walls to shelter their plants and, where the soil was not suitable, they added materials such as fine gravel to improve drainage or blackened the surface with charcoal or added pebbles on the surface to aid the uptake of heat.[1]

The Māori had little need to grow native plants, but sometimes planted a special variety of flax to have it handy (Colenso, 1869a), or a plant of kōwhai-ngutu-kaka (*Clianthus puniceus*) for its handsome red flowers (Allan, 1961). Clarkson (1984) has observed that, even after 150 years, trees apparently cultivated by the Māori, such as karaka, tītoki, and a broad-leaved form of rangiora, remain prominent around abandoned pa sites in Taranaki.[2-4]

4.2 Te Wao Nui a Tane: The National Ethnobotanic Garden

On 21st January 1990, at the Science and Natural Heritage Conference at Massey University, Hon. Peter Tapsell, then Minister of Science, announced that the Pu Hao Rangi Maori Trust of Auckland and DSIR Botany Division would be partners in developing a concept plan for a National Ethnobotanical Garden. The New Zealand 1990 Commission granted $45,000 and the Auckland Regional Council provided land for a trial planting of kūmara, maize, potatoes, flax, gourd, cabbage tree, and Māori medicinal plants. Dell Wihongi of Pu Hao Rangi and Dr Murray Parsons of Botany Division (now DSIR Land Resources, Botany Institute) were commissioned to prepare the concept plan. This will include a plant nursery, herbarium, library, and research, cultural, educational, and tourism facilities.[5]

4.3 The Sweetest Cottage Imaginable

From 1840 onwards, speakers and writers deplored the destruction of native vegetation by land-hungry settlers. The volumes of the *Transactions of the New Zealand Institute* from 1869 on contain a steady flow of papers about the value of the native flora, and the danger of erosion. Among the settlers there were always some who appreciated native bush and retained individual trees or a scrap of the primeval forest.

Mr Justice Chapman, who occupied Homewood, Karori, from 1844 to 1852, cleared only thirty of his 118 acres, and cut a path through the bush. An 1850 watercolour by C.D. Barraud shows a single native tree as a feature of the flower garden around the house.[6] J.B.A. Acland, who settled at the Mount Peel Station in 1856, left an area of beech forest near the homestead, and had a path through it as an extension of the garden. He retained a large cabbage tree as a centrepiece for the flower beds and lawns.[7]

William Swainson, who came to Auckland as Attorney General in 1841, lived at Taurarua, now Judges Bay, where in 1857 he had 'the sweetest cottage imaginable, with lovely tree ferns and native shrubs of all kinds growing about it, and lovely scented roses covering the trellis work in front'.[8]

David Hay, who established the Montpellier Nurseries at Hobson Bay, Auckland, in 1858, listed some natives for sale from his first catalogue. This is not dated; but subsequent issues in 1863, 1865, 1867, and so on all listed some native plants among many exotic kinds.[9]

However, the pioneers were divided in their preferences: James Hay of Pigeon Bay, one of the early Canterbury settlers, observed that an English oak would grow more in ten years than a tōtara would in fifty.[7]

4.4 Baskets, Fenders, Canes, and Whiphandles

In 1870, in the *Transactions of the New Zealand Institute*, T.H. Potts and W. Gray discussed the cultivation and uses of some 150 species of native trees and shrubs. Potts' garden at Governor Bay was their testing ground. For basket-making, they listed bush-lawyer and supplejack, and added that the latter made excellent ship's fenders. They also remarked that supplejack canes would be appreciated in Britain. For screen hedging, they recommended *Coprosma lucida* and *C. robusta, Cordyline australis, Myoporum laetum, Muehlenbeckia complexa, Olearia forsteri* (now *Shawia paniculata*), and *Griselinia littoralis*; and for dwarf edging they nominated *Teucridium parvifolium*. They suggested that the leaves of *Entelea arborescens* be tried as food for silkworms, recorded that the durability and hardness of *Discaria toumatou* made it suitable for whiphandles, and considered that the blossoms of *Rubus* and *Cordyline* would yield commercial perfumes in the future.[10]

4.5 Alpines

One of the first growers of alpine plants in New Zealand must have been J.B.A. Acland, who established Mount Peel station in 1856. He explored the upper reaches of the Rangitata River and collected alpines for his garden.[7]

By 1880, H.J. Matthews, the Dunedin nurseryman, offered twenty-one kinds of *Celmisia*, fifteen of *Olearia*, fourteen of *Senecio*, and fifty kinds of *Hebe* (many of which were probably alpine or subalpine species). He exported plants and seeds to England and other countries (Plate 3).[11]

Since then, gardeners have devised ways of growing almost every kind of alpine plant, using novel arrangements of rocks for shelter and heat, scree for drainage, and peat, pumice, or sand to provide the desired soil conditions. A number of excellent books have been produced on the culture of alpines: in 1945 W.B. Brockie's *New Zealand Alpines in Field and Garden* appeared, with photographs of alpine plants in the Christchurch Botanic Gardens.[12] Twenty years later, in 1965, W.R. Philipson and D. Hearn wrote *Rock Garden Plants of the Southern Alps*, with chapters on scree plants, gentians and ourisias, hebes, southern heaths, native orchids, vegetable sheep and mountain daisies. The superb plates are by D. Hearn, an Associate of the Royal Photographic Society, who contributed a chapter on alpine plant photography.[13] *Growing New Zealand Alpine Plants*, by Joe Cartman, appeared in 1985, with chapters on the natural habitat, cultivation, propagation, pests and diseases, and the plants themselves. Excellent colour and black and white photographs are supplemented by Hugh Wilson's attractive line drawings.[14]

4.6 Carpeters and other ground-cover plants

After the Second World War, increasing contact with the western seaboard of America, and in particular California, brought a change of style to New Zealand gardens. The front lawn, with its surrounds of massed annual flowering plants and rose beds, gave way to the Californian 'pebble garden'. This was a low-maintenance garden where the ground was covered with a layer of small pebbles or stone chips and planted with scattered shrubs.

The novelty of the 'pebble garden' soon wore off because its very simplicity and lack of colour made it monotonous. In the 1960s, gardeners looking for a more acceptable solution to maintenance problems took up the increasing overseas trend to use ground cover plants in order to reduce the cost of weeding and other maintenance. By using a much wider range of plants the gardener's desire for colour, diversity of form and easy maintenance were all satisfied. Among New Zealand plants that have proved of value are *Coprosma kirkii* and similar *Coprosma* cultivars, and *Leptospermum scoparium* low-growing cultivars. In Britain, New Zealand bidi-bids (*Acaena* species), *Hebe* species and cultivars, and *Senecio* 'Dunedin hybrids' and 'Sunshine' have also been used to advantage.[15]

In Australia, the climate is not an easy one for New Zealand plants, but some gardeners have been successful with *Acaena, Cotula, Mazus* and *Nertera* as 'carpeters'; others have had success with *Fuchsia procumbens* and, as noted before, *Raoulia* species, while many have found low-growing *Hebe* cultivars and dwarf *Leptospermums* suitable for keeping out weeds. The experience of one of us (Cooper) is that *Hebe diosmifolia*, planted as a hedge, yields leaf litter that checks the growth of weeds nearby.

Native sedges, grown in New Zealand and overseas for decorative and ground cover purposes include *Carex buchananii*, valued for its reddish-bronze leaf colour; *C. comans*, with a red-leaved variety known as 'Bronze form', and a yellow-green cultivar known as 'Green wig'; *C. petriei*, smaller and again with red leaves; and *Uncinia uncinata*, in a form with bright reddish-brown leaves which has been advertised as the 'Copperware shrub'.

Native grasses are sometimes planted *en masse* to provide variety of form and colour, as well as ground cover, and include toetoe (*Cortaderia species*); broad-leaved snow tussock (*Chionochloa flavescens*); narrow-leaved snow tussock (*C. rigida*); and blue tussock (*Poa colensoi*).

In 1985 Julian Matthews and Gordon Collier described the garden of Dr and Mrs Neil Prentice at Invercargill where native and exotic weed species have been planted to form a 'No Mow' lawn. Exotics used were listed as *Sagina subulata* 'Aurea', and *Thymus lanuginosus*. Native plants were identified as *Hydrocotyle novae-zelandiae, Raoulia lutescens* and *R. glabra*.[16]

4.7 Ferns

The fern craze of Victorian England has been mentioned and fern containers (miniature glasshouses and bottles) are still to be found in some New Zealand homes. The first outdoor fernery in New Zealand is not known. However, H.J. Matthews, a Dunedin nurseryman, offered 100 kinds of ferns for sale in his 1880 catalogue, and he exported plants to England and other countries.[11]

In later chapters we deal with 'Todea fibre' (Chapter 6), hard fern as a ground cover and scent plant (Chapter 11), and ferns as food (Chapter 8).

For lovers of New Zealand ferns, excellent guide books have been available for many years. In 1882, the Dunedin naturalist, G.M. Thomson, wrote *The Ferns and Fern Allies of New Zealand*. Eight years later, in 1890, H.C. Field, engineer, politician, and a successful grower of ferns, wrote *The Ferns of New Zealand*.[17,18] In time their place was taken by H.B. Dobbie's *New Zealand Ferns*. Three editions were revised by Marguerite Crookes, who kept pace with new discoveries and the spate of name changes. After the sixth edition in 1963, Whitcombe and Tombs ceased publishing what had become a classic.[19] The gap was filled by several fern books, among them M. Fisher and L. Ward's *New Zealand Ferns in Your Garden*, 1976, with striking black and white paintings by Mrs Ward.[20] In 1989, *New Zealand Ferns and Allied Plants* by Patrick Brownsey of the National Museum, Wellington, and John C. Smith-Dodsworth of Coromandel, was published. It is an authoritative and well illustrated book, with clear directions for cultivating each kind, and seems likely to please amateur and professional pteridologists for a long time.[21]

4.8 *Azolla*

In Chapter 14, we mention the blue-green alga *Anabaena azollae* (more correctly *Anabaina*), which grows within the water fern *Azolla*.

Overseas many and varied uses have been suggested for *Azolla* some of which have been tried successfully; for example, as a fertilizer for crops, as a food for pigs, rabbits, and fish, and as a decontaminant and scavenger of minerals, potassium, and phosphorus in sewage treatment. On garden and farm ponds, complete surface mats are thought to inhibit the laying of mosquito eggs and to check water evaporation. Research worldwide is reported briefly in the *Azolla Newsletter*, issued quarterly by Dr I. Watanabe, IRRI, P.O. Box 933, Manila, as an international service.[22]

New Zealand studies of possible uses for the indigenous *Azolla filiculoides* and the introduced *A. pinnata* might be worthwhile; they are very common on farm ponds. The old name of *Azolla rubra* was a more descriptive name for the native species.[23]

4.9 Tree Ferns

Whekī (*Dicksonia squarrosa*) was sometimes used to form the outside of better Māori houses, while small pieces of the trunk of mamaku (*Cyathea medullaris*) were placed like a plinth around the base of a house or kūmara store to keep out rats.[2]

On Stewart Island, early European settlers built 'bungi huts' of tree fern trunks, while ponga (*Cyathea dealbata*) huts were built near the edge of the bush in Southland; for example, Mr Howard built a house with walls of tree fern trunks which was later owned and used as a tool shed by Kew Hospital, Invercargill. This and other tree fern houses have been photographed by Sheila Natusch, author of *On the Edge of the Bush*, 1976. In some bush districts, tree fern trunks are still used to form 'corduroy tracks' and, when the ends sprout, there is a border of living fronds along the path.[23]

Tree fern trunks are a feature of New Plymouth gardens, having been used extensively to face banks, hold terraces, and form fences; when the trunks sprout, the newness of the work is hidden in a wealth of young tree ferns. There is a strong demand for the trunks in other centres, but supplies are sometimes erratic.

For many years, Brewsters, a well known New Plymouth nursery firm, made vases and bowls from tree fern trunks, the contrast between the hard black and softer white tissues being much admired. Today, a number of firms supply a variety of articles for gardening (large fibre bowls and plant troughs) and the gift and tourist trade (ashtrays, dishes, bowls, and vases).

In 1985, the Horticultural Market Research Unit of MAF suggested that there were major prospects in Japan for items such as unrooted fern logs which could be grown on again. However, more research is needed, and the Farm Forestry Association may be the only suitable organization to do it. The first and most obvious problem is that tree ferns are large in volume and weight in relation to their value. Possibly growing tips only should be exported. Another difficulty is that steep slopes clothed in stands of tree ferns, particularly in the lower Northland and Auckland districts, mark the site of past slips, and it is unwise to damage the existing cover. Finally, some farmers who have stands of tree ferns prefer to trade stems for the services of casual tradesmen, rather than to tackle the complexities of shipping, quarantine, packing, GST, and so on.

We have dealt with the subject of tree ferns at length to illustrate the many considerations that arise when a new bush product is proposed. We have not mentioned the increasing reluctance of many people to cut bush; recently, there has been increasing pressure for State Forests (now Conservation Department Forests) to be earmarked for recreation; for example, deerstalkers and others want the Tongariro State Forest preserved for sport. A long term solution might be for farm foresters to plant tree ferns as a crop.

Superba Ferns, a native plant nursery at Whenuapai near Auckland, raises ferns from spores to order, mainly for clients in Europe. The most frequent request is for tree ferns, and about 5000 are despatched as 'tube lines' each northern spring.[24]

4.10 Native Orchids for Sale

Horticulture is constantly changing. There seem to be new fashions every spring; for example, during Garden Week at Auckland in 1983, one nursery showed a collection of native orchids and took orders for $15,000 worth of plants. *Calochilus campestris*, classified as 'endangered' by Botany Division, DSIR, was for sale at $12 a plant.[25]

Nine native orchids are described and illustrated in the latest work on rare plants, *Threatened Plants of New Zealand*, by C.M. Wilson and D.R. Given, 1989. Four species are 'endangered' (i.e. in danger of extinction and unlikely to survive), three are 'vulnerable' (i.e. the populations are decreasing because of over-exploitation), and two are presumed extinct.[26] As their habitats are destroyed, the task of saving native orchid species at risk seems insuperable. Edwin D. Hatch of Laingholm, and a few other gardeners have grown them successfully, but growth is slow and some species will not survive in cultivation.[27]

The Native Conservation Council expressed concern at the sale of native orchids from wild populations, and it was hoped that publicity would check the craze. A long-term solution might be research to solve the problems of cultivating native orchids. Also, it must be noted that some orchids offered for sale by nurserymen were collected from the Kaimaumau Swamp before peat mining destroyed the habitat.

For those interested in native orchids, Hatch in 1951 wrote *Auckland's Orchids*, published by the Auckland Botanical Society, 14 Lake Road, Titirangi. A larger work, *Native Orchids of New Zealand*, by John Johns and Brian Molloy, was published in 1983. It describes and illustrates nearly eighty species, with superb photographs of each kind and some shots of the places where they grow.[28]

4.11 Botanic Gardens and Threatened Plants

Botanic Gardens were mooted from the beginning of European settlement. For example in 1840 New Zealand Company directors instructed the Surveyor-General, Lieutenant Mein Smith, to provide a Botanic Garden in the plan of the infant town now known as Wellington City. The early history of the Garden is described in *The Botanic Garden, Wellington, A New Zealand History 1840-1987*, by W. Shepherd and W. Cook (1988).[29]

New Zealand botanic gardens have never been financed by central government on a long term basis; consequently their usual role is to serve as the town or city formal garden. Native plant collections have been made in a haphazard fashion, a notable exception being the Otari Native Plant Museum at Wadestown, Wellington. The attention given to threatened native plants in recent years has changed this.

David Given, Botany Division, DSIR, prepared the original list of threatened native plants in 1976, and updated it several times as more accurate information was obtained. He wrote *Rare and Endangered Plants of New Zealand* in 1981 and, with Catherine Wilson, presented the latest data and illustrations in DSIR Field Guide *Threatened Plants of New Zealand* 1989. The Guide deals with 151 species, of which nine are extinct, or presumed extinct, forty-six are endangered and ninety-six are vulnerable.[25,26]

The gardens now have reasons and incentives for collecting and propagating threatened plants. The long term aims are firstly to save a species from extinction, and secondly to return seedlings or cuttings of it to the wild. The gardens are more likely to receive public support when engaged in this work, have useful information and material for educational activities, and obtain valuable specimens for exchange with other institutions or for sale to keen gardeners. Timaru Gardens has been most active in saving threatened plants and, in 1988, had 100 kinds in cultivation.[30-32]

In 1990, the Forest and Bird Society and the Department of Conservation set up the Threatened Species Trust to handle substantial donations from Comalco and Tasman Forestry for endangered bird management programmes, and it is hoped that similar sponsorships for threatened plants will be forthcoming.[33]

4.12 Plant Breeding

Conditions for breeding new plants have improved; since 1955, the Royal New Zealand Institute of Horticulture has served as the international registration authority for *Hebe* and *Leptospermum* cultivars and the Nomenclature Committee of the Institute has published lists of names, and organized collections of *Hebe* at Otari Native Plant Museum, Wellington, and of *Leptospermum* at Christchurch Botanic Gardens. The University of Canterbury has another collection of *Hebe* at Ilam, the Auckland Regional Council has additional collections of native plant cultivars at the Botanic Gardens, Manurewa, and DSIR Land Resources Botany Institute, Lincoln, has experimental gardens as well as permanent collections of living plants.[34,35]

L.J. Metcalf, Director of Parks, Invercargill City Council, and others have provided 'fancy names', descriptions and illustrations of most recent cultivars of native plants, and breeders now have certain proprietary rights under the Plant Variety Rights Act of 1987. In 1983, G.S. Wratt and H.C. Smith edited *Plant Breeding in New Zealand*, in conjunction with the DSIR, surveying work done on *Agathis, Hebe, Leptospermum* and *Nothofagus*, and recording the names of breeders. They considered that Walter B. Brockie (1897-1972), who received the Plant Raisers Award for *Phormium* 'Smiling Morn' in 1968, was probably our most outstanding breeder of native cultivars, especially of new forms of *Coprosma*.[36-38]

Margaret Jones of New Zealand Flax Hybridisers, Tauranga, is now the foremost breeder of *Phormium* cultivars, and exports plants, and foliage for floral work. Jack Hobbs of the Auckland Regional Botanic Garden has taken up the breeding of *Hebe* cultivars, with the aim of producing forms suitable for the Auckland district. Nurserymen, particularly in the Auckland region, have produced a number of new native plant cultivars; for example, Graham Platt of Albany has selected several, especially in *Metrosideros*. Terry Hatch of Pukekohe has assisted Graham Platt with *Metrosideros*, and also produced other native cultivars (Plate 3).

taupata (*Coprosma repens*) **Rubiaceae**
Engraved from a 1769-70 painting by Sydney Parkinson

Another notable breeder of native plant cultivars is Felix Jury of Tikorangi near Waitara, who has hybridized Celmisias to produce progeny suitable for lowland gardens. He has also worked successfully with *Phormium* to obtain new colour forms, and his crossing of *Astelia nervosa* and *A. chathamica* shows promise.[16]

Apart from those mentioned, few New Zealand gardeners tackle the systematic breeding of native plant cultivars, and the field is wide open for those who wish to take up the challenge.[54]

From 1982 to 1987, the DSIR Climate Laboratory at Palmerston North screened a selection of native plants for frost tolerance. The success of live plant exports is largely dependent on their ability to survive European winters. Hardier native taxa have been identified, and this information should be of assistance to nurserymen in deciding which species and cultivars to export.[39]

PLANT VARIETIES ACT

The N.Z. Coalition for Trade and Development, P.O. Box 11-345, Wellington, has announced its opposition to the Plant Varieties Act, contending that no one has the right to own another living, evolving life form, and that the legislation is helping to erode genetic resources.[40] A somewhat similar view was aired at an International Workshop on Ethnobotany, held at Te Rehua

Marae, Christchurch from 22 to 26 February 1988.[41] The Workshop was organized by Botany Division, DSIR, as part of a Commonwealth Science Council's project on biological diversity and genetic resources. Some Māori participants asked the DSIR to refrain from further experimentation and the Plant Varieties Registrar to refrain from registering varieties derived from New Zealand indigenous plants until DSIR completed thorough consultation with the tangata whenua concerning their wishes about the use of native plants. The claim that the Māori own native plants and information relating to them stimulated discussion and publicity. At a public symposium on New Zealand native plants, entitled '90's Natives', held at the Bay of Plenty Polytechnic from 31 July to 2 August 1990, Dell Wihongi of Pu Hao Rangi alleged that New Zealand scientists were trying to exploit indigenous plants for Pākehā profit. The audience included plant nurserymen, breeders, botanists, and conservationists, many of whom were clearly irritated by the suggestions.[55]

4.13 Levin Horticultural Research Centre

The main New Zealand centre for plant breeding is the Levin Horticultural Research Centre of the Ministry of Agriculture and Fisheries. About 1982, the Levin Centre obtained a number of ecotypes (distinct populations adapted genetically to particular habitats) of kōwhai (*Sophora microphylla*) from Botany Division, DSIR, Lincoln, for propagation and selection of specimen trees and possible pot-plants. Three cultivars, 'Early Gold', 'Goldie's Mantle', and 'Goldilocks' were selected and named in memory of the late G.N.J. Goldie, who established the collection at Levin.[42]

There are large and valuable markets for pot-plants in Europe, USA, and Japan, and even a minor penetration of them would give significant returns to New Zealand. The staff of the Levin Centre developed mini pot-plants for this purpose, and trial shipments of *Leptospermum* to Japan confirmed the potential. Importers from the US, Singapore, and Malaysia also expressed interest.

The Levin Centre tested a number of native plants, and found that the most suitable for pot-plants are: *Asplenium bulbiferum, Blechnum penna-marina, Earina autumnalis* and *E. mucronata, Fuchsia procumbens, Leptospermum scoparium* and cultivars, *Senecio greyi*, and *Sophora* species and cultivars. Problems of mini pots were tackled at Levin, including propagation, control of plant shape and flowering, shelf life, packaging and stringent quarantine regulations in countries overseas. The aim of the work was to produce management blueprints for sale to commercial growers.[43,44]

In 1988, a company named Miniplants International Limited was formed as a joint venture between MAF Techmark Services and DFC Ventures, MAF taking 40% and Agricola Resources Limited taking the balance of capital.

However, for the year to 30 June 1989, Agricola reported losses on its other ventures totalling 7.2 million dollars. Three staff at Levin working on the miniplants were dismissed, and it was announced that the New Zealand market is too small, quarantine restrictions overseas are too difficult, and in future, Miniplants International Ltd will license growers in other countries.[45]

4.14 Hebes

In 1989, Linda N. Kristensen, a Danish scientist from the Institute of Glasshouse Crops at Aarsley, collaborated with Dr Julie Plimmer of the Department of Horticultural Science, Massey University, and Dr Ian Warrington of Plant Physiology Division, DSIR, Palmerston North, in growth-chamber experiments on some fifty species, 200 clones and 150 cultivars of *Hebe*. In Denmark, annual sales of a New Zealand cultivar known as *Hebe x franciscana* total about 1,500,000 pots, worth about N.Z. $15,000,000. Linda Kristensen is looking for other Hebes suitable for the Danish market. Preliminary results from the growth-chamber experiments indicate that, before they will flower, high-altitude *Hebe* species need far more exposure to cold than low-altitude species.[46]

In 1989, Dr Warwick Harris, then Director of Botany Division, suggested in an article in *Commercial Horticulture* that New Zealand nurserymen should have an advantage as this country is the centre of the *Hebe* genetic resource. Efforts should be made to locate and protect remaining *Hebe* populations in the wild. Communication and cooperation between growers and scientists should be the aim. He urged growers to use the names *Hebe* and *Veronica* correctly, to ignore pressure to market plants as Veronicas, and to try the Māori name 'koromiko' for promoting sales.[47]

4.15 Cut Foliage

Chris Barnaby, a graduate in horticultural science at Massey University, Palmerston North, in 1989 surveyed the markets for cut foliage, and found that Bay of Plenty, Taranaki, and Auckland growers, most of whom are cultivating less than a hectare of land, are supplying local and overseas markets with considerable quantities of cut foliage. Both exotic and native plant foliage is grown: the natives including *Leptospermum*, *Lophomyrtus*, *Phormium*, and *Pittosporum*. Overseas markets exist in Japan, Europe, and USA, especially for unusual-looking cut foliage.[48]

4.16 New Zealand Gardening Books

Michael Murphy, author of the *Handbook of Gardening for New Zealand*, first published by Whitcombe and Tombs in 1885, included a chapter on native plants in the fourth edition of 1907. Leonard Cockayne assisted with the list, which included trees, shrubs, climbing plants, herbaceous and alpine plants, and plants for a sub-tropical effect.[49]

In 1911, H.H. Travers' eleven-page booklet *Cultivation of Native Plants* was printed by *Otago Daily Times*, Dunedin. It contained notes on eighty-seven genera, and a list of more than 260 species and varieties, classed as climbers and border, shrubbery and rockery plants. He suggested that readers who wished to grow the choicest alpines look at what Mr W.H. Taylor had achieved in a shade house and well-drained rockery and bank at the Experimental Farm at Weraroa. As he had not had the advantage of the Government chest or the Government Printer for the issue of his booklet, he trusted the public would purchase it freely, and that it would be as useful as *New Zealand Plants and their Story* by L. Cockayne, which had been printed by the Government Printer![50,51]

Cockayne, as early as 1914, wrote *New Zealand Plants Suitable for North American Gardens*. In 1924, his series of articles on native plants suitable for the British Isles was published and, in the same year, his most effective work on *The cultivation of New Zealand Plants* appeared.[52,53]

A popular work at present is *Gardening with New Zealand Plants, Shrubs and Trees*, by Muriel E. Fisher, E. Satchell, and Janet M. Watkins, first published in 1970. The book has chapters on the role of native plants in ornamental planting, coastal gardens, school grounds, screens and hedges, rock and alpine gardens, the fernery, orchid house, pebble garden, bonsai, and water gardens. However, the most authoritative work for the serious grower of native plants is *The Cultivation of New Zealand Trees and Shrubs* by L.J. Metcalf, a revised edition of which was produced by Reed/Methuen in 1987. Not only does Metcalf cover propagation, cultivation, pests and diseases, as well as the suitability of native plants for different situations, but he provides descriptions and illustrations, and for New Zealand cultivars his book is probably the best reference work available.[54,55]

1. Leach, H. *1000 Years of Gardening in New Zealand* Wellington (Reed) 1984 157p.
2. Colenso, W. (1869a, 2nd ed. 1875) 1 *Trans. N.Z. Inst.*, 233.
3. Allan, H.H. *Flora of New Zealand Volume 1* Wellington (Government Printer) 1961 1085p.
4. Clarkson, B. (1984) *Botany Division Newsletter*, Suppl. 1, p. 14.
5. Parsons, M. (1990) 130 *Botany Division Newsletter* 6.
6. Smedley, B. *Homewood and its Families. A Story of Wellington* Wellington (Mallinson Rendel) 1980 163p.
7. Strongman, T. *The Gardens of Canterbury. A History* Wellington (Reed) 1984 203p.
8. Reed, A.H. (ed.) *Further Maoriland Adventures of J.W. & E. Stack* Dunedin (Reed) 1938 258p.

9. Challenger, S. (1985) 13 *Annual Journal Royal N.Z. Inst. Hort.* 60. (1986/7) 14, ibid. 76.
10. Potts, T.H., Gray, W. (1870) 3 *Trans. N.Z. Inst.* 181.
11. Gow, P.B. (1986/7) 14 *Annual Journal Royal N.Z. Inst. Hort.* 86.
12. Brockie, W.B. *New Zealand Alpines in Field and Garden* Christchurch (Caxton Press) 1945 32p.
13. Philipson, W.R., Hearn, D. *Rock Garden Plants of the Southern Alps* Christchurch (Caxton Press) 1st ed. 1962, 2nd ed. 1965, 167p.
14. Cartman, J. *Growing New Zealand Alpine Plants* Auckland (Reed Methuen) 1985 106p.
15. Thomas, G.S. *Plants for Ground Cover* London (Dent) 1970 273p.
16. Matthews, J., Collier, G. *The New Zealander's Garden* Auckland (Endeavour Press) 1985 135p.
17. Thomson, G.M. *The Ferns and Fern Allies of New Zealand* Dunedin 1882 164p.
18. Field, H.C. *The Ferns of New Zealand* Wanganui 1890 164p.
19. Dobbie, H.B. *New Zealand Ferns* Auckland (Whitcombe & Tombs) 2nd ed. 1921, etc. (4th, 1931; 5th, 1952; 6th, 1963; revised by Marguerite Crookes).
20. Fisher, M.E., Ward, L. *New Zealand Ferns in Your Garden* Auckland (Collins) 1976 111p.
21. Brownsey, P.J., Smith-Dodsworth, J.C. *New Zealand Ferns and Allied Plants* Auckland (Bateman) 1989 168p.
22. Watanabe, I. (ed.) (1986) 2(1) *Azolla Newsletter* 1, and subsequent issues.
23. Natusch, S. *On the Edge of the Bush* Wellington (Craig Printing) 1976 115p.
24. Vandermast, S., Superba Ferns, 80 Trig Road, Whenuapai, Kumeu R.D. 2 — personal commun.
25. Given, D.R. *Rare and Endangered Plants of New Zealand* Wellington (Reed) 1981 154p.
26. Wilson, C.M., Given, D.R. *Threatened Plants of New Zealand* Wellington (DSIR Publishing) 1989 151p.
27. Hatch, E.D. (1955) 12(3) *Auckland Bot. Soc. Newsletter* 4, June. 12(4) ibid. 3, September.
28. Johns, J., Molloy, B. *Native Orchids of New Zealand* Wellington (Reed) 1983 124p.
29. Shepherd, W., Cook, W. *The Botanic Garden Wellington. A New Zealand History 1840-1987* Wellington (Millwood Press) 1989 396p.
30. Given, D.R. (1986/7) 14 *Annual Journal Royal N.Z. Inst. Hort.* 3. 1986 Banks Lecture on Botanic Gardens and Conservation.
31. Heenan, P.B. (1986/7) 14 *Annual Journal Royal N.Z. Inst. Hort.* 18.
32. Given, D.R. (1988) 249 *Forest & Bird* 25.
33. Smith, K. (1990) 256 *Forest & Bird* 10.
34. Nomenclature committee of the Royal N. Z. Institute of Horticulture (1963) 'Checklist of *Leptospermum* cultivars'. 5(5) *New Zealand Plants & Gardens* 224.
35. Nomenclature Committee of the Royal N.Z. Institute of Horticulture (1970) 'Checklist of *Hebe* cultivars'. 2(1) *J. Roy. N.Z. Inst. Hort.* 25.
36. Metcalf, L.J. (1963) 5(3) *New Zealand Plants & Gardens* 120.
37. Metcalf, L.J. (1969) 1(4) *J. Roy. N.Z. Inst. Hort.* 172.
38. Wratt, G.S., Smith, H.C. (eds) *Plant Breeding in New Zealand* Wellington (Butterworths with DSIR) 1983 309p.
39. New Zealand DSIR (1988) *Plant Physiology Division Biennial Report* 1988, p. 19.
40. Bell, A., [n.d.] *A growing matter. An inventory of the vegetable and native plant*

seeds of Aotearoa/New Zealand Wellington. (N.Z. Coalition for Trade and Development).

41. Harris, W., Kapoor, P. (eds) *Nga Mahi Maori O Te Wao Nui A Tane, Contributions to an International Workshop on Ethnobotany.* Christchurch (Botany Division, DSIR). 1990 209p.

42. Butcher, S.M., Wood, S.M.N. (1985) 13 *Annual Journal Royal N.Z. Inst. Hort.* 52.

43. New Zealand Ministry of Agriculture and Fisheries, Agricultural Research Division. *Annual Report for 1982-1983,* Levin Horticultural Research Centre, p.165.

44. Butcher, S.M. MAF Levin Hort. Res. Centre, personal commun. 25 Jan 1985.

45. *New Zealand Herald,* 10 January 1990. 'Plant venture uprooted . . .'.

46. Warrington, I. (1989) 27 *DSIR Biological Industries Group Newsletter* 3.

47. Harris, W. (1989) *Commercial Horticulture,* August, p. 40.

48. *Waitomo News,* 9 November 1989, p. 22 'Cut foliage'.

49. Murphy, M. *Gardening in New Zealand* Christchurch (Whitcombe & Tombs) 1907, 4th ed., 311p.

50. Travers, H.H. *The Cultivation of Native Plants* Dunedin (*Otago Daily Times*) [1911] 11p.

51. Cockayne, L. *The Cultivation of New Zealand Plants* Christchurch (Whitcombe & Tombs) 1924a 139p.

52. Cockayne, L. *New Zealand Plants Suitable for North American Gardens.* Panama Pacific International Exposition, San Francisco, USA, February-December 1915, Wellington (Government Printer) 1914 35p.

53. Cockayne, L. (1924b) 88 *The Garden* 615, 632, 646, 660.

54. Fisher, M.E., Satchell, E., Watkins, J.M. *Gardening with New Zealand Plants, Shrubs and Trees* Auckland (Collins) 1978 303p.

55. Metcalf, L.J. *The Cultivation of New Zealand Trees and Shrubs,* Auckland (Reed Methuen) 1987. Revised ed. 346p.

56. Metcalf, L.J., 117 Moana Street, Invercargill. Letter of 21 November 1990.

57. Garnock-Jones, P. (1990) 21 *N.Z. Botanical Society Newsletter* 14.

5

From the Hills to the Seashore

5.1 Māori Use of Timber

In the previous version of this book (Brooker, Cambie, and Cooper 1988)[1] readers were referred to the writings of authors such as Colenso and Best for information regarding traditional Māori use of wood. From these sources it seemed that the Māori preferred tōtara (*Podocarpus totara*) for large carvings, canoes, and the framing of their houses, although north of Thames they used kauri (*Agathis australis*). Mataī (*Prumnopitys taxifolia*) and pukatea (*Laurelia novae-zelandiae*) were also chosen for carving, while mairetawhake (*Syzygium maire*), maire (*Nestegis cunninghamii*), ake (*Dodonaea viscosa*) and mānuka (*Leptospermum scoparium*) or kānuka (*Kunzea ericoides*) provided tough woods for implements and weapons. Straight stems of mānuka and tawa (*Beilschmiedia tawa*) served as battens in house-building, and lengths of kareao (*Ripogonum scandens*) were used to tie pieces of framework together and to hold thatch in place.[2-4]

We remarked that it would be of interest to check the identity of woods in museum collections of Māori artefacts, using the services of a competent wood technologist. Since 1967, R.N. Patel, Botany Division, DSIR, has published a series of papers on the anatomy of indigenous conifer and dicotyledon wood. In 1978, B.A. Meylan, Physics Engineering Laboratory, DSIR, and B.G. Butterfield, Botany Department, University of Canterbury, Christchurch, completed DSIR Bulletin 222 containing 805 scanning electron micrographs showing the structure of 115 New Zealand woods. The two lines of research provided a basis for Rod Wallace, an Otago-trained archaeologist, to identify the woods used in Māori wooden artefacts from Museum collections.[5,6]

In his first paper, Wallace (1982) identified eight conifer and four hardwood species used in forty-four adze hafts and twelve adze sockets. The artefacts belonged to six collections. A small slip of wood was removed from each artefact for sectioning and light-microscope identification. All the hafts and sockets were made from branches of trees rather than from the more familiar trunk wood. From the results it was clear that the old-time Māori preferred light flexible wood for hafts and heavy wood for sockets. The results did not agree with the ethnographic records, indicating that these records only poorly reflect the prehistoric situation.[7]

Wallace (1989) discussed the wood used in 762 Māori artefacts from five museums.[8] Results from these papers and more recent work are summarized below:-

Wood	Māori Use
1. kauri (*Agathis australis*) trunkwood	carvings, house timbers, canoes, wakahuia.
branch wood	fern root beaters, mallets, tapa beaters, weapons, one piece spades.
2. mataī (*Prumnopitys taxifolia*) trunkwood	carvings, bowls, house timbers, wakahuia, spade handles.
branch wood	fern root beaters, mauls, adze handles, spinning tops.
3. tōtara (*Podocarpus totara*) trunkwood	carvings, house timbers, canoes, bowls.
branch wood	adze handles, beaters.
4. rimu (*Dacrydium cupressinum*) tanekaha (*Phyllocladus trichomanoides*) kawaka (*Libocedrus plumosa*) Hall's tōtara (*Podocarpus hallii*) kahikatea (*Dacrycarpus dacrydioides*) trunkwood	house timbers, bowls, canoes.
branch wood	fern root beaters, adze handles.
5. 'māpara' or 'kāpara'*	spinning tops, hair combs.
6. mānuka (*Leptospermum scoparium*), kānuka (*Kunzea ericoides*), maire (*Nestegis* sp.), rātā (*Metrosideros* sp.), pōhutukawa (*Metrosideros excelsa*), pūriri (*Vitex lucens*), kōwhai (*Sophora microphylla*), akeake (*Dodonaea viscosa*).	fernroot beaters, mauls, paddles, weapons, spade blades, weeders, digging sticks, bird spears (kānuka and maire).

7. māpou (*Myrsine* sp.), porokaiwhiria (*Hedycarya arborea*), ramarama (*Lophomyrtus bullata*), and other small broad-leaf trees	digging sticks, composite adze handle sockets (māpou)
8. tawa (*Beilschmiedia tawa*)	light house timbers, paddles, battens.
9. pukatea (*Laurelia novae-zelandiae*)	bowls, palisade posts.

* resinous heart of rimu and kahikatea, found in the centre of rotten logs in the bush.

It is notable that unlike modern commercial timber use, Māori exploitation of wood involved branch wood and other parts of the tree. This wood often differs from trunkwood in its properties. Kauri branchwood, for example, is heavily resin-impregnated and is very hard, heavy, and tough.

Wallace (pers. comm. 1990) also reports that the study of a collection of wooden artefacts from a wet archaeological site is showing evidence of a prehistoric exchange network involving wood. The site is located south of the kauri line (the southern limit of kauri is *c.* 38 degrees south) but contains significant amounts of this species used in making artefacts.

5.2 Logging and Sawmilling

European knowledge of native timber trees was first recorded in October 1769, when Captain James Cook and his men spent several days at Tolaga Bay near the East Cape of the North Island of New Zealand. Here they examined the shores and filled the casks with good water which was found conveniently close to the beach. Upwards of twenty different kinds of trees were cut down, and Mr Banks and Dr Solander obtained samples of their wood, and branches with flowers and fruit which they described, drew, and made into dried specimens.[9]

Accounts of Cook's voyages gave publicity to the timber trees of New Zealand, especially those at Dusky Bay and at the River Thames (now the Waihou River). In *Murihiku and the Southern Islands* (1907), Robert McNab recorded the history of European exploitation and settlement of Fiordland and the Southern Islands from 1770 to 1829. In *New Zealand Forests* (1984) Harriet Fleet provided an account of similar exploitation and settlement for the whole country. Of particular interest are the records relating to the Waihou River; for example, in March 1795, the brig *Fancy* returned to Sydney with the first cargo of spars from that locality. The crew spent three months gathering two hundred trees from sixty to one hundred and forty feet in length.[10,11]

The superiority of kauri timber was soon realized, and traders searched the northern coast of the North Island for supplies of spars. The trade in kauri logs was set back by the burning of the *Boyd* in 1809; however, after the *Dromedary* obtained a full load of spars from Whangaroa Harbour in 1820 the industry developed rapidly.

5.3 Thomas Kirk and the Kauri

Thomas Kirk, while Professor of Natural Sciences at Wellington College (a teaching affiliate of the University of New Zealand), was instrumental in persuading the Premier, Julius Vogel, to bring down the first New Zealand Forests Act of 1874. Captain Inches Campbell Walker, Madras Forest Department, was appointed first Conservator of Forests, and Kirk accompanied him on part of the tour Walker made soon after his arrival, and introduced him to the Colony's forest flora.

Unfortunately Walker left in 1876, and forest destruction continued without restraint until 1885, when Vogel brought down a second Forests Bill. Under this Act, Kirk was appointed Chief Conservator of Forests, but the Stout–Vogel Government was defeated in 1887, and the new Atkinson Ministry retrenched rigorously, including abolishing the Forestry Branch. By the time it was disbanded early in 1888, Kirk had 800,000 acres of forest reserves set aside and had nearly finished an illustrated account of native forest trees. This book, *The Forest Flora of New Zealand*, was published in 1889, and although 100 years have passed, many of Kirk's comments are relevant today; for example, of kauri he warned that exhaustion was not far distant.[12,13]

Northern kauri forests once covered 1.5 million ha, but less than 4% remains. Milling reached a peak in 1906, when 440,000 m^3 of kauri was cut, but cutting alone did not cause destruction; at least half of the kauri forest was burnt to clear the land for grass. Some fires, such as those in the Puhipuhi forest in the 1880s, burnt for months.[14]

From 1921, when the NZ Forest Service was formed, there was an official policy to acquire land on which young second-growth kauri was growing, and this was endorsed in the 1973 Government Kauri Policy. However there were never adequate funds for the policy, and land in kauri second growth was cleared for farming or planting in radiata pine.

For many years Forest Service officers carried out research with a view to perpetuating kauri, both in natural stands and in managed forests; they planted many thousands of kauri seedlings in selected areas, and gave sometimes controversial 'silvicultural treatment' to free them from other plants. However, losses from the 'Kauri Estate' far exceeded these replantings; for example, Omahuta Forest kauris were 'butchered' in the 1940s for wartime boat-building; Warawara Forest was milled from 1964 to 1973 'to relieve unemployment', until cutting was stopped by public protests; Puketi Forest was milled until 1979, when cutting was suspended to protect kōkako, and

Herekino Forest was largely cleared of kauri over the same years.

When the 1983 Kauri Management Review was undertaken by the Forest Service, it attracted 776 public submissions, almost all opposing the Service's plans and seeking more reserves, but none of these submissions was accepted.[15,16]

By 1980 it was clear that removal of kauri trees from virgin forests must end. Future kauri timber supplies would have to come from regenerating forests and from artificially established plantings. Both kinds of stand would need intensive management to provide an acceptable financial return. One suggestion was that new planting should be a mixture with suitable exotics, which would act as a nurse crop and yield an intermediate return.[17]

5.4 Professor McGregor and Waipoua

In about 1948, Professor W.R. McGregor of Auckland University College led a successful campaign to prevent the logging of mature trees in Waipoua Kauri Forest, and a struggle has continued ever since to maintain Waipoua Kauri Sanctuary.[18] Maintaining the public road through the forest without damaging the trees has been a problem, made more serious when widening and sealing of the road was begun, to cope with increasing numbers of tourists. However, more destructive than the effects of tourism are the possums browsing on forest plants. The Department of Conservation has been forced to undertake aerial drops of poison on a large scale to reduce possum numbers.

Professor McGregor suggested that Waipoua Kauri Forest should be part of a Kauri National Park, and others have suggested that the sanctuary be extended to the coast. In 1985, the New Zealand Native Forest Restoration Trust, assisted by Forest and Bird and the Queen Elizabeth II National Trust, successfully raised funds to add over 146 ha of cut-over forest to the Sanctuary, the new area to be known as the W.R. McGregor Reserve.[19]

In 1988, the Northland National Parks and Reserves Board proposed a National Park for the North along the lines suggested forty years before by Professor McGregor. The proposal has been studied and endorsed by the National Parks and Reserves Authority, the Northern Regional Council, Tai Tokerau District Maori Council, and the Department of Conservation.[20,21]

The Whangarei Office of DOC advertised in July 1990 for submissions from the public regarding a Kauri National Park, and tāngata whenua are being consulted at appropriate hui. The proposed National Park will comprise up to thirty-five areas totalling about 90,000 ha of forest containing kauri.

5.5 The National Forest Survey and FRI

In 1945 A.R. Entrican, Director of Forests, instituted the National Forest Survey to ascertain the distribution of species, possible timber yields, and erosion potential of our remaining forests. The survey was planned by A. Priestley Thomson (North Island) and J.T. Holloway (South Island), using aerial photographs with regular ground sampling by a permanent staff of twenty, augmented by students and forest trainees in summer. The results of the survey, published in 1956, showed that little millable native forest remained. Much of the remaining forest was needed to limit erosion and to reduce excessive run-off and flooding, and could never be milled. The survey also made evident the damage being caused by introduced animals, and highlighted the excessive waste of much milling.

The Forest Research Institute was set up in 1947 to handle the survey results. Staff working on production forestry and forest products are located at Rotorua, while staff in the Protection Forestry Division are based in Christchurch. Since 1987, FRI has formed a major part of the new Ministry of Forestry, which was established to provide information services to the Government and to industry.[22]

Following the survey, the Forest Service set aside further sanctuaries and ecological areas totalling nearly 375,000 ha, as samples of all the important forest and scrubland communities in New Zealand. However, on 4 July 1975, a number of conservation groups met at Maruia, where they signed a charter of unity, formed the 'Native Forests Action Group' and started a huge petition to Parliament seeking the 'conservation of native forests'.

5.6 An End to Logging Native Forests?

In 1971 the Forest Service issued a White Paper on 'The Utilization of South Island Beech Forests', and followed it in 1973 with a booklet on *Beech Forests* by A. Kirkland, Director of Forest Management. The booklet was illustrated with attractive photographs by J.H. Johns, and explained the beech management plans to the general public.[23,24] Both publications produced more opposition than support; the most common objection was to the milling of native beech trees and their possible replacement with radiata pine. As many of the proposed beech production areas were in forest previously set aside for 'protection purposes' (i.e. erosion control), considerable concern was expressed by conservation groups and interested laypeople.[11]

As early as 1973, there had been strong public protests at the clear felling of native forests between Tauranga and Rotorua, referred to as the Bay of Plenty or Kaimai-Mamaku forests.[25] Then conservationists began to agitate for the preservation of virgin forest remnants between Te Kuiti and Taupo. In 1978 the Forest Service received more than 1400 submissions, most seeking a stop to milling at Pureora and, after public protest which included conservationists sitting in trees that were about to be felled, the Government announced a moratorium on milling for three years and the payment of $7 million to three milling companies. However this did not solve the problem of Waihaha, another 'virgin forest remnant', where milling was underway.[26-28]

In 1981, the Native Forests Action Group as the 'Joint Campaign on Native Forests' pinpointed 'Selection Logging' as a cause of major damage in forests by wind-throw of exposed trees; the following year the Group targeted woodchip mills at Awarua near Invercargill and at Richmond near Nelson, criticizing wasteful destruction of native forest remnants and the payment of regional subsidies. The Forestry Encouragement Grant Scheme, which frequently resulted in the clearance of native bush for the planting of radiata pine, was attacked in 1983; the Group's attractively illustrated paper *Bush Telegraph* was distributed widely to the press, politicians, and conservationists. In the same year, the Group publicized planning studies of the Ministry of Works and Development which showed that radiata pine plantings at the East Cape were only marginally economic and at Marlborough Sounds were uneconomic.[29-32]

By the time of the 1984 general election, the Labour Party had promised to protect all remaining North Island native forests; however, owing to the number of people employed in the Westland timber industry, nothing was offered for the South Island. There, beech logging was the cause of most controversy, with lack of reserves a major concern; for example, when the Forest Service published the North Westland Management Plan in 1981, with proposals for clear felling and burning large areas of native forest, more than 5000 submissions were made, urging larger reserves, wildlife corridors, and an end to destruction. Similarly, the 1982 Buller Management Plan aroused 2600 submissions, most seeking an end to bush clearance; but, as with the North Westland plan, the Forest Service saw no merit in the submissions.[33]

It was inevitable in the circumstances that the thrust of the Group's representations should turn to strengthening the conservation sector of the Government administration of native forests. Late in 1985, the Government announced that a Forestry Corporation would be set up to manage commercial production forests, and a Department of Conservation would be formed to administer all forests not being used for wood production, all Crown Lands not being used for agriculture, national parks, reserves, foreshores, coastal waters, wild and scenic rivers, wildlife, and historic places.[34]

In early November 1986 a committee under the chairmanship of Dr Roger Blakeley, Ministry for the Environment, negotiated an agreement to end the fifteen years of agitation for more adequate South Island native forest reserves. The 'Forest Accord', between the various conservation interests, West Coast United Council, West Coast Sawmillers' Association, and Timber Workers' Union, provided for a network of fifty-eight new reserves covering 137,000 ha, and a new national park. With about 2400 people dependent on logging and sawmilling, some concessions had to be made; rimu logging would continue

Plate 1

puawānanga (*Clematis paniculata*) **Ranunculaceae** (p.30)
Painting: E. Blumhardt

Plate 2

kahikatea (*Dacrycarpus dacrydioides*) **Podocarpaceae** (p.11)
Photo: J.E. Braggins

parrot's bill (*Clianthus puniceus*) **Fabaceae** (p.27)
Photo: L.H. Briggs

Plate 3

tikumu (*Celmisia*) (p.48)
Painting: E. Blumhardt

pōhutukawa (*Metrosideros excelsa*) **Myrtaceae** (p.53)
Photo: L.H. Briggs

Plate 4

wharariki (*Phormium cookianum*) **Phormiaceae** (p.37)
Painting: F. Osborne

for at least three years in production-zoned forest, and 77,000 ha were set aside for a beech production scheme.[35]

5.7 Te Waipounamu

After a conservation campaign which lasted more than twenty years, and which was intensified from 1985 onwards, the government agreed in February 1989 that 311,000 ha of rainforest between Fox Glacier and Jackson Bay should be protected. This cleared the way for the Department of Conservation, the Ngai Tahu Trust Board, and the Royal New Zealand Forest and Bird Protection Society to prepare a document nominating 2.6 million ha of South Westland for World Heritage status. This vast wilderness area, now known as Te Waipounamu, includes mountains, lakes, fiords, rivers, four national parks, large areas of lowland podocarp forest, beech forest, tussock grassland, and wetlands of world interest.

A rain-forest visitor centre is to be opened at Haast, with co-operation from local people, to provide natural history programmes for the increasing numbers of tourists.[36,37]

5.8 The Tasman Accord

In June 1989 Tasman Forestry, a subsidiary of the Fletcher Challenge Group, agreed to abandon plans for logging fifty-two areas located in various parts of New Zealand from Coromandel to Otago and totalling 38,599 ha of native forest. An agreement known as the 'Tasman Accord' was signed on 22 June 1989 by Tasman Forestry, the Minister of Conservation, the Forest and Bird Society, the Federated Mountain Clubs and the Maruia Society. Tasman's new policy included leaving native forests undisturbed as gene pools, and obtaining timber from plantation forests.[39]

On 25 July 1990 Fletcher Challenge announced a gift of $200,000 to improve public access to privately-owned native forests, and to aid public enjoyment of them.[38]

5.9 The Native Timber Export Ban

In 1989 more than 4000 people, including members of groups and firms, made submissions to the Minister of Conservation regarding the logging of native forests and the export of native wood chips. Only a few submissions were in favour of logging. The Prime Minister, Hon. Geoffrey Palmer, announced in March 1990 that the export of sawn native timber, logs, and woodchips would be banned except where production is from an area of native forest managed under a certified sustained yield plan, or where sawn native timber is surplus to domestic needs.

Discussion since March 1990 has been on the effects of the export ban. Both politicians and conservationists have expressed the wish for a complete ban on logging remaining areas of native bush. For example, in May 1990 the Minister of Forests, Hon. J. Sutton, told a Christchurch conference of farm foresters that he, personally, would like to see the end of felling of indigenous forests except where the owner agreed to 're-establishment with native species'.[40,41]

At the 1990 Tree Conference held at the University of Auckland in June 1990, conservationists resolved as part of a Tree Manifesto, 'that the utilization of indigenous wood should be from trees specifically planted and cultivated, not from intrusion into public heritage forests'.[42]

Farmers who own blocks of native bush have expressed concern at the losses they are suffering, and logging contractors in some localities, especially western Southland, claim that their livelihood is threatened. The Ministry of Forestry has advised affected landowners and contractors that they may lodge claims for compensation. Farmers with native forest 'of ecological value' are urged to covenant it to the Queen Elizabeth II National Trust. The Trust can help the farmer to fence the forest and to obtain rate relief from the local authority.[43]

The National Government's forest policy 'to end uncertainty' was announced by the new Minister of Conservation and Minister for the Environment on 21 December 1990. The new prescription for logging native forest is that the rotation period must be 20% longer than the age of the fully mature forest. This is to ensure that there is always some mature forest left. In native podocarp forest and virgin beech forest, only single trees or small groups may be logged, and low impact techniques must be used. The Government has also set aside $6.75 million dollars in the Forest Heritage Fund to buy high value forests on private land, and $3.35 million a year in Nge Whenua Ranui, a separate funding scheme to buy forests on Māori land. There is to be a substantial transition period during which cutting will continue 'above the sustainable level'. Conservationists immediately criticized the resumption of woodchipping in Southland beech forest, and the new policy seems likely to provoke bitter opposition from them.[65]

5.10 Native Forest Reserves

Bush reserves administered by the former Lands Department or Forest Service were sometimes not fenced, and funds to control animal pests, eliminate weeds, and maintain tracks were seldom adequate. The new Department of Conservation has similar funding difficulties, although its role is better defined.

At Trounson Kauri Forest near Donnellys Crossing, the trees were damaged by wind until a protective belt of fast-growing exotic conifers was planted. The trees in Sterling Kauri Bush near Matakohe need a similar protective belt. As well as for protection of the native forest remnant, fast-growing exotics may be planted to provide an income for essential maintenance. H. McKellar (1990) suggested that where a part of a reserve is in old grassland or low scrub, this could be planted in radiata pine. The planting should be accessible, so that it can be logged later without damage to the native forest. It should be managed adequately to obtain the best possible return. McKellar's suggestion will be rejected by many people, but all too often when a reserve is established the funding for its management is quite inadequate.[44]

5.11 Revegetating with Native Trees

From 1959 to 1961 the Forest Research Institute at Rotorua undertook a massive replanting of partially-logged native forest on the Mamaku Plateau, using 33,000 seedling rimu, tōtara, kahikatea, matai and tanekaha. Logging had removed nearly all the rimu, and the gaps left had become filled with hardwood shrubs, tree ferns, and toetoe.

A bulldozer was used to open lanes, clear gaps, and pile up the rubbish into windrows. After fifteen years the strips of original forest between the lanes were devastated by wind and dieback. Some of the seedlings had failed to grow in soil compressed by the bulldozer, others had been smothered by surrounding vegetation.

After twenty-five years, survival rates of the seedling rimu, tōtara and kahikatea were between 45% and 70%, of matai 50% and of tanekaha 47%. The experiment showed the need to rip up compacted ground before planting, to choose good sites for the seedlings, and to release them from surrounding vegetation at regular intervals.

The authors of the report, D. Bergin, G. Pardy and T. Beveridge (1988) suggested that fast-growing exotic conifers such as radiata pine might be used on cleared land to provide shelter and an interim income. They considered that a better result would be obtained on a site with a more favourable soil and climate. They concluded that the cost of replanting logged native forest is very high, with no economic return for many years.[45]

The New Zealand Native Forest Restoration Trust, formed after the Pureora tree-sitting episode, has replanted cleared and damaged forest at Pureora, at Cowan's Wildlife Reserve, and at Rangitoto Station in the Rangitoto Range. The Trust uses volunteer labour for planting and has raised more than 100,000 native tree seedlings at an Oratia nursery.

The replanting of Tiritiri Matangi Island in the Hauraki Gulf to make an 'open' bird sanctuary has been carried out by volunteers, organized by a resident ranger who is a member of the staff of the Department of Conservation. Seedling trees are raised in a nursery on the island. The project has had considerable publicity, and there is a waiting list for those who wish to visit the island.

However, revegetation with native trees will be limited as long as the cost of native seedlings is far higher than the cost of radiata pine seedlings. Aerial distribution of native tree seed and slow-release fertilizer has been suggested to reduce costs. It is necessary to find some method to hold the mix on bare ground, and carpets of straw and bark and fascines of cut mānuka have been tried.

For information on planning, obtaining young trees, and managing a revegetation project, reference should be made to the *Revegetation Manual* of the Queen Elizabeth II National Trust.[46-48]

5.12 Farm Forestry

Now that the logging of native forests has almost ended and many timber mills seem to be equipped to handle only radiata pine, the future of native timber as a product seems bleak. However, John Millet, in the February 1987 issue of the *New Zealand Tree Grower* argued the merits of tōtara (*Podocarpus totara*) as a desirable farm tree. At present there are no significant plantings of tōtara although there is a looming shortage of special-purpose and decorative timber for manufacturing; and supplies of logs for Māori carving are inadequate. John Millet listed the tree's advantages as wind tolerant, useful for stabilizing steep banks, stock proof and so able to protect other native plants except under heavy browsing. The great drawback to planting tōtara for timber is the length of time it takes to grow, compared with exotic trees.[49]

In the same journal for August 1987, David Bergin and George Pardy of FRI discussed growing tōtara as a plantation crop. They stressed the need to choose good sites for seedlings, and to tend them by adequate pruning and by release from surrounding vegetation. They concluded that tōtara may have potential for growing in small woodlots as a future timber source; however, further research is needed on wood quality of fast-growing trees and on factors which influence heartwood formation.[50]

George Pardy and David Bergin were the authors of another paper on native trees in the February 1989 issue of *Tree Grower*. This paper reported the preliminary research results from raising and planting six kinds of native trees; tītoki (*Alectryon excelsus*), kohekohe (*Dysoxylum spectabile*), rewarewa (*Knightia excelsa*), pukatea (*Laurelia novae-zelandiae*), mangeao (*Litsea calicaris*), and pūriri (*Vitex lucens*). The authors discussed collecting and preparing seed, germination, raising seedlings, undercutting taproots, root pruning, and preliminary results from planting-out trials. One objective of the research is to find native hardwood trees that can replace increasingly expensive imported timbers as an alternative to locally-grown exotic species.[51]

In the *Tree Grower* for August 1990 Bergin and Pardy discussed the results of a survey of native tree plantings in fifty-five localities in the North and South Islands. Of native conifers, tōtara (*Podocarpus totara*), kauri (*Agathis australis*), kahikatea (*Dacrycarpus dacrydioides*) and rimu (*Dacrydium cupressinum*) were planted most frequently. Pūriri (*Vitex lucens*) was the most common hardwood tree, occurring in one third of stands, followed by pōhutukawa (*Metrosideros excelsa*), karaka (*Corynocarpus laevigatus*), tītoki (*Alectryon excelsus*) and rewarewa (*Knightia excelsa*) in the North Island, and red beech (*Nothofagus fusca*), black beech (*N. solandri* var. *solandri*), and silver beech (*N. menziesii*) in the South Island. Only six plantations had been tended adequately; in the other forty-nine, young trees had not been released from smothering vegetation, and thinning and pruning had not been carried out.

The main reason for planting native trees was given as beautification. Other reasons given were to augment forest remnants, to revegetate cleared areas, to provide food and shelter for birds, to protect erosion-prone sites, to provide a future wood supply, to obtain better shelter, and to enhance the natural heritage.[50]

In the first version of this book[1] we suggested that farm foresters might look at the possibility of planting or retaining small areas of native trees for other purposes.

The first plot might be cabbage trees (*Cordyline* species) for fructose production. It is improbable that the extraction of *Cordyline* fructose will be economically viable compared to that of starch-derived fructose, but as a 'natural' novelty and native 'additive' it could be a market success.

A second planting might be of tree ferns. A Wanganui farmer found a market for tree fern fibre as a garden mulch and suggested that there were possibilities in Australia for the product.

The third lot could be planted in mānuka for charcoal. Again, a Wanganui farmer found a market for the charcoal with those who like charred chops cooked on a barbeque.

5.13 Wet Bottoms

Bogs, swamps, and oxbows along a stream or river help to protect land lower down and provide special habitats for animals and plants. Reduction in these features through draining and 'reclamation' (using landfill) diminishes their capacity to hold and cleanse silt-laden floodwater.

In 1958 there were floods at Te Kuiti and Otorohanga, and stopbanks were raised along the Waipa River at Otorohanga. Since then, bush on the Rangitoto Range has been cleared in many places, bogs along the riverbanks have been drained, run-off of rainwater has been speeded up and the margin of flood protection provided by the Otorohanga stopbanks has been reduced.

In 1989, the New Zealand Native Forest Restoration Trust appealed for $250,000 to buy the Rangitoto Station at the headwaters of the Waipa River. Members have begun replanting native bush to check erosion on the Station and to slow water run-off. However, bogs, swamps, and oxbows are needed down-stream below the station, as they are needed down the banks of most other similar rivers.

5.14 Salt Marshes and Mangrove Swamps

Plants of salt marsh which were of value to the Māori people included oioi (*Leptocarpus similis*), sea-rush (*Cyperus ustulatus*), and rīrīwaka (*Bolboschoenus fluviatilis, B. medianus,* and *B. caldwellii*); all were used for the outer thatch of houses. Oioi was the best material on account of its durability. The function of different rushes and sedges in mat-making are noted in Chapter 6, and the use of *Bolboschoenus* tubers as an emergency food is noted in Chapter 8.

Of salt marsh plants, *Sarcocornia quinqueflora* (formerly *Salicornia australis*) has the common name 'glasswort' referring to the former use of ashes of a similar European plant in glass-making, and also the common name 'samphire' (*herbe de St Pierre*), referring to the use of similar plants overseas in salads and pickles.

In his 1869 paper on Economic Botany, William Colenso recorded that woody stems of mangroves (*Avicennia resinifera*) were used for making soap, and the bark for tanning.[4] More recently, a number of authors (Knox 1980, Crisp 1986, for example) have discussed the productivity of mangrove swamps and salt marshes. Professor George Knox, formerly of the University of Canterbury, Christchurch, gave estuarine 'productivity' as 2000g of plant matter per square metre a year, compared with 730g for land and 155g for the ocean. Pam Crisp (of the Nature Conservation Council, Wellington) gave mangrove litter 'production' as eight tonnes per hectare a year, making the mangrove swamp one of the highest litter producers of all New Zealand forest types. Crisp pointed out that at least thirty commercially important species of fish, such as flounder and mullet, use estuaries at some stage in their life cycles, and that disturbance or pollution could have severe effects on fish stocks.[52-54]

Professor N.H. Nickerson, Tufts University, Medford, USA, who spent sabbatical leaves in New Zealand in 1979-80 and in 1987, considered that stock grazing on mangroves prevented fruiting and removed up to half of the mangrove litter, to the detriment of fish stocks. He recommended (1980, report on Rangaunu Harbour) that farmlands be fenced to keep stock from mangrove swamps. Nickerson (1980, report on mangroves to the ARA Parks Department) commented on research at the Leigh Marine Station which indicated that mangroves 'fix' atmospheric nitrogen for their own use in building protein. In the same paper he discussed the role of bacteria, in the anaerobic muds of mangrove swamps and salt marshes, in breaking down nitrates to release atmospheric nitrogen. This latter process cleanses the water of tidal inlets, removes excess nutrients, and has a significant effect on the quality of coastal water.[55-58]

Hicks and Silvester (1985) reported that samples of estuarine muds at two New Zealand localities contained active nitrogen-fixing bacteria. Activity is low, and is increased in the presence of plants, especially in the presence of decaying litter. Most nitrogen fixation in New Zealand mangrove swamps is probably carried out by epiphytic algae on the mangrove pneumatophores.[59]

5.15 Buffer Strips

Nickerson (1980, report on Parengarenga Habour) suggested that the mangrove area of the harbour was diminishing and not replacing itself. He gave evidence that the shellfish resource was dwindling. He blamed heavy grazing of mangroves by both cattle and sheep, and gave instances of stock on the mud-flats. He also blamed the silt, sand, and mud (from apparently overstocked farmland) for the loss of both mangroves and shellfish.[57]

In 1987, Nickerson found that several of his 1980 recommendations had been adopted by one farm manager. Fences had been moved back a full chain from tide lines to provide a buffer zone between farmland and seashore, stock had been fenced from gullies, wide bands of scrub and bush had been left between pastures on newly broken in land, and fencing had been started to keep stock from the mangroves. Freed from the twin pressures of grazing and root-asphyxiation as sand and clay covered pneumatophores, mangroves had made vigorous growth, and produced many seedlings. These had rooted in a new, fine, soft silt layer, apparently organic, that covered the formerly sterile tidal flats below the farm block. Nickerson provided some evidence that numbers of small fish in the mangroves, and of shellfish on the flats, had increased since his first visit in 1979-80.[58]

Nickerson (1987, report on southernmost mangroves) found mangroves in Raglan Harbour at one locality only, Waingaro Landing, where a population of about 1000 plants was backed by a steep bank covered in native bush. Elsewhere in the harbour the tidal flats are bare of mangroves. The hills around the harbour are essentially bare of bush, except along the northern rim near Raglan north head; here there are still remnants of bush, developed for holiday baches. Below, at about high-tide mark, are stumps of mangroves, apparently cut out to improve the flats.

Nickerson urged again that a buffer zone be provided between land and sea, to protect mangroves from mud washing down from farmland. He pointed again to the need for fences to reduce the grazing of mangroves by stock. He argued that restoration of mangroves and salt marshes would result in an increase in the number of shellfish and young fish feeding on organic materials provided by the vegetation, cleaner water, and better fishing in the long run.[58]

Early in 1990, several residents of Kawhia circulated a petition urging the Government to ban trawlers from fishing off the coast at Aotea Harbour. They claimed that overfishing had resulted in fewer snapper surviving to maturity. However, efforts to persuade Parliament to ban trawlers have generally been unsuccessful; if all coasts thought to be overfished were closed, there would be few trawlers at sea.[60]

It might help future fishing if residents sought the planting of mangroves in the Kawhia, Aotea, and Raglan Harbours, with buffer zones of trees and shrubs on land behind the mangroves. The buffer strips might be extended up streams entering the harbours to help reduce the inflow of silt from erosion in the hills behind.

Norman and Alison Fletcher, in a letter to the *New Zealand Herald* dated 20 August 1990, described damage caused to Whangapoua Harbour by the inflow of silt following 1971 bush clearance of the Whangapoua Forest. No buffer strips were left along stream edges, and a heavy fall of rain caused erosion, flooding, and a slick that extended miles out to sea. The Whangapoua Harbour has never recovered.[61]

Similar tales could be related about other North Island harbours which once had extensive, undamaged mangrove forests and excellent fishing.

In Chapter 14, we suggest the need for a buffer zone of vegetation at Lake Omapere, and in Chapter 15 we mention the provision of buffer zones along steams carrying nutrients into Lake Taupo.

5.16 Sand-dune Plants

As the European settlement of New Zealand proceeded, portions of fertile coastal farms were sometimes buried under advancing sand dunes, especially on the west coast of the North Island. At a meeting of the Auckland Institute in 1873, James Stewart, Civil Engineer, cited examples of sand encroachment between Port Waikato north head and Waiuku, and between Muriwai and Kaipara south head. Thomas Kirk recommended beginning the work of reclamation as near as possible to high-water mark, to slow the entry of new sand. He considered that marram grass (*Ammophila arenaria*) from Europe was the best pioneer sand reclamation plant, but he suggested that native sand fescue (*Austrofestuca littoralis*), pīngao (*Desmoschoenus spiralis*), silvery sand grass (*Spinifex sericeus*), and tussock grass (*Poa caepitosa*) should be tried at or near the strand.

Kirk listed other native plants that he thought would prove to be valuable

marram grass (*Ammophila arenaria*) **Poaceae**

lupin (*Lupinus arboreus*) **Fabaceae**

pine (*Pinus radiata*) **Pinaceae**

Three introduced plants used to stabilize dunes on the Aupouri Peninsula, 1980.
Photos: N.H. Nickerson

akeake (*Dodonea viscosa*) **Sapindaceae**
Engraved from a 1769-70 painting by Sydney Parkinson

sand-stabilizers further inland, including toetoe (*Cortaderia* species), sea rush (*Cyperus ustulatus*), sea spurge (*Euphorbia glauca*), cottonwood (*Cassinia* species), karo (*Pittosporum crassifolium*), and akeake (*Dodonaea viscosa*).[62]

In 1909 and 1911, Leonard Cockayne provided the Government with two classic reports on the sand dunes of New Zealand; he favoured planting marram on the foredune and in loose sand, then tree lupin (*Lupinus arboreus*) to improve fertility, and finally *Pinus radiata* for long term cover and to provide a cash crop. He suggested also that toetoe could be a valuable sand stabilizer, but it has not been tried on a large scale; in fact, adventive species of *Cortaderia* are regarded as a pest and a fire risk in some radiata pine forests.[63,64]

There have been few changes in the list of plants recommended for sand reclamation since Cockayne furnished his reports more than seventy-five years ago. Fore-dunes covered in native *Spinifex* are now rare, pīngao and sea spurge are classified as 'vulnerable species', and akeake-covered dunes are almost unknown. Some people deplore the 'excessive planting' of marram, lupin, and radiata pine; most people are not aware of what has been lost.

1. Brooker, S.G., Cambie, R.C., Cooper, R.C. *Economic Native Plants of New Zealand* Christchurch (Botany Division, DSIR) 1988 130p.
2. Best, E. *The Maori As He Was* Wellington (Dominion Museum) 1934 280p.
3. Best, E. *Forest Lore of the Maori* Wellington (Government Printer) 1942, reprinted 1977, 420p; and other works by Best, E.
4. Colenso, W. (1869, 2nd ed. 1875) 1 *Trans N.Z. Inst.* 233.
5. Patel, R.N. (1988) 26 *N.Z.J. Bot.* 337, and earlier papers.
6. Meylan, B.A., Butterfield, B.G. *The Structure of New Zealand Woods* Wellington (DSIR Science Information Div.) 1978. DSIR Bulletin 222, 250p.
7. Wallace, R.T. (1982) 4 *N.Z.J. Archaeology* 179.
8. Wallace, R.T. 'A preliminary study of wood types used in pre-European Maori wooden artefacts' in Sutton, D.G. (ed.) *Saying So Doesn't Make It So. Papers in Honour of B. Foss Leach* (N.Z. Archaeological Ass. Monograph 17) 1989 p. 222-32.
9. Beaglehole, J.C. (ed.) *The Journals of Captain James Cook* 4 vols., printed for the Hakluyt Society by Cambridge University Press. Vol. 1 *The Voyage of the Endeavour 1768-1771*, 1955. 2nd ed. 1968 684p.
10. McNab, R. *Murihiku and the Southern Islands* Invercargill (W. Smith) 1907 377p.
11. Fleet, H. *New Zealand's Forests* Auckland (Heinemann) 1984 175p.
12. Brown, L. *The Forestry Era of Professor Thomas Kirk F.L.S.* Wellington (N.Z. Forest Service) 1968 25p.
13. Kirk, T. *The Forest Flora of New Zealand* Wellington (Government Printer) 1889 345p. 142 pl.
14. Adams, G.E. *Kauri, A King among Kings* Auckland (Wilson & Horton) 1977, revised ed. 81p.
15. Anon. (1984) 14 *Bush Telegraph* 4.
16. Anon. (1985) 20 *Bush Telegraph* 7.
17. Barton, I.L., Horgan, G.P. (1980) 25 *N.Z.J. Forestry* 199.
18. McGregor, W.R. *The Waipoua Kauri Forest* Auckland 1948 80p.
19. Bellingham, M. (1986) 241 *Forest & Bird* 16.
20. Ell, G. (1988) 248 *Forest & Bird* 4.
21. Anon. (1988) 47 *Forest & Bird Conservation News* 2.
22. *Forest Research Institute Report* 1987, Wellington (Ministry of Forestry) 71p.
23. N.Z. Forest Service (1971) 'White Paper on the Utilisation of South Island Beech Forests', in *Report of the Director-General of Forests for the Year ended 31 March* 1971, p. 12.
24. Kirkland, A. *Beech Forests* Wellington (N.Z. Forest Service) 1973 48p.
25. Anon. (1983) 10 *Bush Telegraph* 1; 11 ibid. 4; 12 ibid. 2; 13 ibid. 1.
26. Collingwood, D. (1978) 209 *Forest & Bird* 25; 210 ibid. 22.
27. Forest Service (1981) 221 *Forest & Bird* 8.
28. Edmonds, A. (1982) 223 *Forest & Bird* 14.
29. Anon. (1981) 1 *Bush Telegraph* 6.
30. Anon. (1982) 7 *Bush Telegraph* 8. 9 *ibid* 4.
31. Anon. (1983) 12 *Bush Telegraph* 4. 13 ibid. 3.
32. Anon. (1983) 13 *Bush Telegraph* 6.
33. Anon. (1984) 15 *Bush Telegraph* 1.
34. Anon. (1985) 22 *Bush Telegraph* 1.
35. Anon. (1987) 27 *Bush Telegraph* 6.
36. Anon. (1989) 252 *Forest & Bird* 19. 254 ibid. 1 (editorial by A. Mark).
37. *Forest & Bird Annual Report* 1990, p. 3.

38. *New Zealand Herald* 26 July 1990, Section 1, p. 13 'Conservation projects to boost forest access'.
39. McSweeney, G. (1989) 253 *Forest & Bird* 10.
40. *New Zealand Herald* 13 March 1990 'Clamp on native wood export'.
41. Sutton, J. (1990) 11(2) *New Zealand Tree Grower* 5.
42. Anon. (1990) 3(5) *Growing Today* 10.
43. *The New Zealand Farmer* 29 August 1990, p. 7 'Kneecap jolt for timber owners'.
44. McKellar, H. (1990) 11(2) *New Zealand Tree Grower* 21.
45. Bergin, D., Pardy, G., Beveridge, T. (1988) 9(2) *New Zealand Tree Grower* 44.
46. Brown, B., Heath, B. (1986/87) 14 *Annual Journal Royal N.Z. Inst. Hort.* 41.
47. Lee, M. (1987) 246 *Forest & Bird* 28.
48. Evans, B. *Revegetation Manual. A Guide to Revegetation Using New Zealand Native Plants* Wellington (Queen Elizabeth II National Trust) 1983.
49. Millet, J. (1987) 8(1) *New Zealand Tree Grower* 14.
50. Bergin, D., Pardy, G. (1987) 8(3) *New Zealand Tree Grower* 68, (1990) 11(3) ibid. 2.
51. Pardy, G., Bergin, D. (1989) 10(1) *New Zealand Tree Grower* 3.
52. Crisp, P. *Coastal Wetlands* Wellington (Nature Conservation Council) 1986. Information Booklet No. 25 16p.
53. Gregorie, D. (1980) 218 *Forest & Bird* 2.
54. Knox, G.A. (1980) 16(2) *Soil & Water* 13.
55. Nickerson, N.H. *Freshwater Wetlands, their nature and importance to man* Medford (Lincoln Filene Center, Tufts University) 1978 8p.
56. Nickerson, N.H. (1980a) 'Auckland Regional Parks — A Series of Reports Made to the Manager, Parks and Reserves'.
57. Nickerson, N.H. (1980b) A Series of Reports on Ecological Conditions . . . made to the Assistant Director-General of Lands . . .'.
58. Nickerson, N.H. (1987) 'A Series of Reports Reviewing Changes since 1979-80'.
59. Hicks, B.J., Silvester, W.B. (1985) 49(4) *Applied & Environmental Microbiology* 955.
60. *Waitomo News* 22 February 1990, no. 884, p. 1.
61. *New Zealand Herald* 20 August 1990. Letter to the Editor by N. & A. Fletcher.
62. Kirk, T. (1874) 6 *Trans. N.Z. Inst.* 45.
63. Cockayne, L. *Report on the Sand Dunes of New Zealand. The Geology and Botany, with their Economic Bearing* Wellington (Government Printer) 1909 30p.
64. Cockayne, L. *Report on the Dune-Areas of New Zealand, their Geology, Botany and Reclamation* Wellington (Government Printer) 1911. 74p, Published with 63 as Appendices to the Journals of the House of Representatives, C13 1909: 1-30, and C13 1911: 1-74.
65. *New Zealand Herald*, 17 April 1991 'Beech exports approved'.

6

Fibres and Paper

6.1 New Zealand Flax

In former times, the Māori people used flax (harakeke) as their principal fibre plant, and recognized a number of named varieties for special purposes. They split selected green leaves to make kits (for food-gathering, cooking, and storage), nets, mats, cloaks, sandals, ropes, carrying-straps, and even sieves (lined with toetoe tops) to extract the poisonous seeds of tutu berries. On the north bank of the Mokau River, a short distance above the bridge, there are grooved stones scattered for many metres, suggesting the vast size of a flax fishing-net which finally rotted there.

Dressed and dyed strips of flax were made into decorative patterned kits and tukutuku panels — a craft that is flourishing again today. The silky white fibre that was obtained by hand-scraping was superior to any produced later by mechanical means, and the fine mats and cloaks made from it are treasured today as works of art and rarely seen outside museums.[1,2]

From about 1820, the Māori prepared increasing quantities of flax fibre for export and sometimes took payment in guns and ammunition to wage war on their enemies. The export industry flourished for about forty years and then dwindled. It was revived in the late 1860s, with the invention of scraping and percussion machines to produce fibre by mechanical means. By 1870, exports had increased again to a maximum value of $264,000.[3]

A report by James Hector in 1872, to the Commissioner appointed by the Government to investigate the flax industry in New Zealand, contained chemical and microscopical studies by William Skey. In addition to investigating methods for the preparation of fibre, Skey also reported on the bitter and gummy materials of flax, and on the effects of acids, alkalis, and organic solvents on the leaves.[4]

The main centre of activity at this time was the Manawatu, where, in the early 1900s, 20,000 acres of flax plants were growing along the banks of the Manawatu River, and where local mills were producing 14,000 tonnes of flax each year. The peak year was 1916, when 27,500 tonnes of fibre valued at *c*. $6 million were exported from New Zealand. Unfortunately, the trade dwindled again in the face of competition from the manila and sisal industries which, despite having a lower productivity, were better organized.[5]

During the Depression of the early 1930s, the Government attempted to rejuvenate the industry by setting up a flax wool-pack industry at Levin. Scientists of the DSIR were appointed to assist with the flax-stripping process, and to establish good quality control. Attempts were also made to develop flax as a source of cellulose, and investigations were carried out on the production of ethanol by fermentation of the juice, and on the use of stripper waste as a fertilizer. Despite work by Maclaurin and Donovan of the DSIR on the possibility of using chemical methods (for example, potassium permanganate) for bleaching the fibre, the traditional method of natural bleaching on fences was never displaced (Plate 6).[6,7]

Yellow-leaf disease, which caused the death of considerable areas of flax in the Manawatu, was traced to a virus carried by a leaf-hopper; control methods were devised (by flooding), and resistant plants were selected. However the flax industry dwindled again, and exists today on a vastly reduced scale. During World War II, attempts were made to use the fibre as reinforcing in fibrous plaster wallboards, but 'stained areas sometimes appeared on surfaces due to water-soluble substances that were left in incompletely washed and milled fibres'.[8,43]

More detailed accounts of the trade in New Zealand flax fibre, later offered as 'New Zealand hemp' in an effort to boost sales, will be found in Critchfield (1951) and Sparrow (1965), while the statistics to 1939 have been traced by Prichard (1970).[9-11] In June 1986, following a number of fatal motor accidents in Auckland and Wellington, it was announced that flax, toetoe, and pampas grass would be studied by the National Roads Board as possible median strip 'barriers' on the country's motorways; however to date, only steel and concrete barriers have been installed.[44]

Today, the most common craft product made from flax is the kit (kete), usually plaited by a Māori woman and sold in a tourist or gift shop. It is useful for carrying shellfish from the beach, and for collecting vegetables and weeds from the garden and pūwhā for the hāngī. Kits were popular with students for carrying books, but weak handles were a nuisance. In one or two districts, handles were taken around the bottom of the kit for extra strength, but usually they are attached to the top of the bag and tend to fail when it is loaded. In 1984 Gillian Elmslie of Mangawhai made much larger containers, and although these were popular as works of art, she had difficulty obtaining enough leaf for her work.[12,13]

From 1986 to 1988, Geoff Walls of Botany Division, DSIR, led a pilot study of the traditional uses of plants in New Zealand and the tropical Pacific as part of the Biological Diversity and Genetic Resources Project of the Commonwealth Science Council. Plants selected for special study included New Zealand flax, cabbage trees, and pīngao. The principal achievement of the pilot study was the establishment of a National Flax Collection in September 1989. About sixty varieties of flax, many of which had been gathered by Rene Ormiston of Gisborne over a period of thirty years, were planted in the collection, which is duplicated with one set at Havelock North in the North Island, and the other set at Lincoln in the South Island. Further varieties will be added to the collection as they are found. Information from Māori experts and from library and museum sources has been assembled as a data base for the collection.[14]

6.2 Phormium Overseas

During the 1790s, the British authorities decided to establish a flax industry on Norfolk Island, where *Phormium tenax* is a native plant. Two Māori men were taken to the island to advise on processing the leaves, but were of little help as this work was done by Māori women. David Collins, Secretary-General of the Colony of New South Wales, wrote an account of the Norfolk Island venture, and it was published in London in 1798.[15]

A. Thouin (1803) wrote of the desirability of introducing 'le lin de la Nouvelle-Zélande' into cultivation in France. With his account of the first flowering in France, Faujas de Saint Fond (1812) published a translation of Collins's account of the Norfolk Island manufactory, a note on the removal of gum from flax leaves, and M. de La Billardière's figures for the strength of the fibre. It is possible that the need to cultivate flax for the French navy diminished with the ending of the Napoleonic Wars in 1815.[16,17]

In Ireland, *Phormium tenax* was repeatedly planted last century for manufacturing purposes, but a successful industry was never developed. Fred Moore, Curator, Royal Botanic Gardens, Glasnevin, in a report for the Lord Lieutenant (2nd September 1885), wrote that it required the best soil, heavy manuring, and was often killed or damaged during severe winters.[18]

New Zealand flax is an important crop on the island of St Helena; plantings increased from 8 ha in 1870 to 1288 ha in 1964. It is grown in wetter localities over 4500 m above sea level, and 70% of the production is exported to the United Kingdom. Most of the remainder goes to Scandinavia and South Africa, frequently mixed with other types of hemp.[19,20]

Other countries to grow New Zealand flax successfully are Argentina, Australia, Brazil, Chile, the Congo, Italy, Japan, Kenya, Mexico, South Africa, Tanzania, and the USA (California). In Brazil, pulp left after the fibre has been extracted is fermented to produce alcohol, and the residue is then fed to cattle.

6.3 Other Fibres

The Māori people in former times employed plants other than harakeke (New Zealand flax) for fibre; Jeanne Goulding compiled a fresh list of these plants in 1971, and cited reference specimens in the Auckland War Memorial Museum's collections.[21]

The leaves of kiekie (*Freycinetia banksii*) were utilized not only for coarse garments, but for baskets, belts, panels, and floor mats or sleeping mats.[21] In January 1989, the *New Zealand Herald* reported that Forest Research Institute, Rotorua, was undertaking research, partly funded by the Department of Conservation, into supplies and harvesting of kiekie. Increased quantities of kiekie are needed for tukutuku panels in new meeting houses and in older houses that are being renovated. Weavers also use the leaves in mat- and container-making, and sales of kiekie articles to tourists have increased. Harvesting needs attention; pulling the leaves off the stems does less damage than cutting the stems down, and may stimulate new growth.[22]

The leaves of tī rākau (*Cordyline australis*) (Plate 6), tī ngahere (*C. banksii*), and tī tōī (*C. indivisa*) were used for many purposes: baskets, bird snares, cordage, rain capes, sandals, and thatching.[21] David Monro MD, in an essay on the Geographical Botany of the Provinces of Nelson and Marlborough for the New Zealand Exhibition of 1865, recorded that Māori sandals made of cabbage-tree leaves were tougher and longer-lasting when green leaves were used. He also noted that, in Otago, the epidermis from the underside of leaves of *Celmisia coriacea* was twisted into a string and used for fishing lines, woven leggings, and moccasins.[23]

Colenso (1869a) stressed the value to the Māori of the broad leaves of tī tōī; they are available in wetter mountain areas where harakeke is not always available, and they were plaited to make strong heavy rain capes (tōī), dyed black and greatly prized. The midrib provided a fine, red-coloured strip for plaiting. Ropes made from tī tōī leaves were very strong, and lasted longer in water than those made from flax. The value of this cabbage tree for craft work is obvious, and it might be worthwhile to cultivate it in suitable localities.[24]

The leathery leaves of pīngao (*Desmoschoenus spiralis*) dry a bright yellow, and are often seen in patterned baskets and tukutuku panels. This sedge was common on coastal dunes, but it has been replaced on many beaches by the introduced marram grass. Māori Arts and Crafts Groups have tried to care for existing wild stands, and to cultivate the plant, but it is not easy to raise from seed or to transplant. Officers of Botany Division, DSIR, attended weavers' huis at Te Teko in 1984 and Ngauruawahia in 1985, to supply biological information about the sedge.[25]

On 21 January 1989, the *New Zealand Herald* reported that the Forest Research Institute, Rotorua, was setting up a major experimental planting of

Plate 5

miro (*Prumnopitys ferruginea*) **Podocarpaceae** (p.103)
Painting: E. Blumhardt

Plate 6

Drying racks of N.Z. flax (*Phormium tenax*) at Foxton about 1940 (p.80)
Photo: J.E. Braggins

tī rākau (*Cordyline australis*) **Asphodelaceae** (p.82)
Photo: R.C. Cambie

Plate 7

toetoe (*Cortaderia*) **Poaceae** (p.84)
Photo: I.A. Williamson, Ninety Mile Beach, 1984

ear fungus (*Auricularia polytricha*) Waitakere Range (p.109)
Photo: P. Buchanan

Plate 8

supplejack (*Ripogonum scandens*) **Smilacaceae** (p.103)
Painting: E. Blumhardt

pīngao 'to restock the plant in its native habitat and to identify guidelines for the rehabilitation of dune lands'.[26]

Subsequently, in March 1990, *DSIR Resource Research* announced that Botany Division DSIR and the Department of Conservation had succeeded in developing techniques for growing pīngao from cuttings and from seed: 'It will now be possible to set up pīngao gardens to supply weavers with all the fibre they want, and to increase wild pīngao with nursery grown plants.' The need to maintain and encourage the growth of wild pīngao was stressed. This requires fencing of wild populations to keep out beach buggies, browsing animals and heavy-footed people.[27]

The Māori names 'kūkuta', and 'kutakuta' seem to refer to two different members of the sedge family, *Scirpus lacustris* (also known as 'paopao' and 'wawa'), and *Eleocharis sphacelata*. In October 1984, a group of Māori weavers from Te Rarawa Marae, Pukepoto, went to the Māori and Pacific Island Weavers' Hui at Te Teko, wearing 'kuta' hats, and attracted much attention. Many people present asked about the material, but supplies were said to be limited. Then the Tai Tokerau District Council and other groups expressed concern to the Mangonui County Council regarding plans to spray 'kuta' at Lake Ngatu to make more space for swimmers. According to a 1984 Koha television programme, 'kuta' used at Pukepoto to make hats and mats is *Scirpus lacustris*, but one of the authors (Cooper) saw Māori women at Ahipara, near Pukepoto, gathering 'kutakuta' (*Eleocharis sphacelata*) from deep drainage ditches, and later admired soft mats made from it. Jeanne Goulding checked collections in the Auckland Museum and found baskets, mats, and rain capes made of *Eleocharis*; there were no articles made of *Scirpus*.

At another display of weaving, in Northland, there were hats made of lacebark (*Hoheria* species), and they were admired for their softness and light weight. Augustus Hamilton (1896) recorded that houhere was formerly made into thin bands as a head dress, and listed similar soft fibres of whauwhi from *Plagianthus*; autetauranga from *Pimelea*; aute from the introduced *Broussonetia*; and even sea-grass (*Zostera*) which was made into a cap (pōtae tauā) which was worn by widows in former times.[28]

The Māori employed raupō (*Typha orientalis*) extensively in house-building, particularly for walls and internal linings, and Goulding listed its use in making sails, fishing rafts and floats, and kites. Nīkau (*Rhopalostylis sapida*) was utilized for lining roofs; according to Goulding, it was also used to wrap rats for cooking, and to make baskets and floor mats. Leaves of toetoe (*Cortaderia* species) were used for lining the walls and roof of a house, and for making baskets, kites, and mats. The leaves were also plaited into cylindrical containers (tūkohu) for steeping food and cooking it in a hot spring. Stalks of toetoe are a part of the framework of tukutuku panels (Plate 7).

In Tuhoeland in former times, the fronds of mauku (*Asplenium bulbiferum*) were woven into a mat which was worn as a covering at night. It was too frail to use outside during the day. Another fern, mangemange (*Lygodium articulatum*), has a frond that grows to a length of 30m or more. This was utilized to make creels and fish traps, and to tie timbers, walls, and thatch together when building huts.[29]

Of the fern allies, Allan Cunningham (1836-1839) recorded the practice of wearing waewaekoukou (*Lycopodium volubile*) as a head 'adornment', while Hamilton noted it as a sign of mourning. He also listed its use in floor mats, and in a shoulder cape probably from the Taupo district. More recently, fibre of Prince of Wales Feather fern (*Leptopteris superba*) was sold for orchid-growing as 'Todea fibre' but, as the fern is rare, the use of fibres from sedges (*Carex*) and tree ferns (*Dicksonia* and *Cyathea*) has been recommended.[30,31]

Finally, mention must be made of the visit to New Zealand in February 1985 of Dr M. Sadahira and Mr T. Akagi, Mat Rush Branch, Hiroshima Agricultural Experiment Station, Japan, who collected seeds of New Zealand rushes (*Juncus* species), and are raising them at Fukuyama for a mat-rush breeding programme with the cosmopolitan *Juncus effusus*.[32]

6.4 Paper-making

Efforts to develop a paper-making industry in New Zealand utilizing indigenous materials were described by Goulding (1964). Some of the poorer quality flax fibre shipped to England was used in paper-making as early as the 1830s, and John Murray produced an account of *Phormium tenax*, printed on paper made from its leaves, at the Morton Paper Mills, Yorkshire, in 1836.[33,34]

At the New Zealand Exhibition, Dunedin, in 1865, there were samples of paper made for John Murray's book of 1836, 'half stuff' made from *Phormium* by Robert Cameron, Christchurch, and paper made from snow grass by Edward McGlashan of Dunedin. The jurors considered that coarse brown paper made from flax fibre would be eminently suitable for bank notes, but there has been no rush by the Reserve Bank to adopt it.[35]

The jurors also suggested that further experiments be made with native grasses, cabbage tree, nīkau, kiekie and astelia leaves, and lacebark fibre. Subsequently, doubts were expressed as to whether New Zealand had sufficient raw material and, in 1873, Thomas Kirk pointed out in a lecture to the Auckland Institute that great quantities of suitable plant fibres were destroyed each year with the progress of settlement.[36,37]

In about 1876 the Mataura Paper mill opened, using native tussock grass from eleven acres beside the Mataura River, as well as flax fibre, rags, sacks, and old rope. The Woodhaugh Mill, which opened on the banks of the Leith Stream, Dunedin, about the same time, also used flax fibre and waste material.

Both mills were stimulated by the offer of a grant of 1500 pounds by the Central Government, and both made coarse paper suitable for paper bags and wrapping. The tussock, which proved to be too brittle, caused some problems with paper quality.[38,39]

Thomas Kirk (1889) suggested that kahikatea (*Dacrycarpus dacrydioides*) could be obtained in vast quantities, and might provide material for an important paper-making industry. However, from about 1885, the trees were milled to make parchment-lined butter boxes until boxes gave way to cardboard cartons in the 1940s. Much timber was exported to Australia for the same purpose, and now only 2% of the dense stands remain.[40, 41]

In recent years, research has shown that flax (*Phormium*), tōwai (*Weinmannia*) and tawa (*Beilschmiedia*) could be used for paper-making; however supplies of them are inadequate, and our paper industry is based on introduced radiata pine. A few native plants are employed by craftspeople in paper-making, but quantities used are minute. May B. Davis, author of an attractive little book on the art of paper-making, listed reeds and sedges as suitable. Unfortunately, she did not identify them further.[42]

1. Pendergrast, M. *Raranga Whakairo. Maori Plaiting Patterns* (Coromandel Press) 1984a.
2. Pendergrast, M. *Feathers and Fibre. A Study of Traditional and Contemporary Craft* Auckland (Penguin) 1984b 237p.
   Pendergrast, M. *Te Wahi Kete — Maori Basketry for Beginners* Auckland (Reed Methuen) 1986 revised ed. 68p.
   Pendergrast, M. *Te Aho Tapu — The Sacred Thread* Photographs by Brian Brake. Auckland (Reed Methuen) 1987a 124p.
   Pendergrast, M. *Fun with Flax — 30 Projects for Beginners* Auckland (Reed Methuen) 1987b 111p.
3. Hutton, F.W. *New Zealand Flax* (Auckland Institute) 1870 17p.
4. Hector, J. *Phormium tenax as a Fibrous Plant* Wellington (Colonial Museum) 1872 134p, 1889 2nd ed. 95p.
5. Joiner, W.A. 'Chemistry in the Phormium (New Zealand Flax) Industry' In *Chemistry in the Development of New Zealand Industry* Wellington (N.Z. Institute of Chemistry) 1940 68p.
6. Hughson, W.G.M., Ellis, A.J. *A History of the Chemistry Division* Wellington (N.Z. DSIR) 1981 201p.
7. Maclaurin, J.S. (1940) 21B *N.Z.J. Sci. Tech.* 263. (Summary compiled by Donovan, W. from Maclaurin's unpublished notes).
8. Boyce, S.D. et al. (1953) A34 *N.Z.J. Sci. Tech. Suppl* 1, 41.
9. Critchfield, H.J. (1951) 5 *Economic Botany* 172.
10. Sparrow, C.J. (1965) 41 *Economic Geography* 331.
11. Prichard, M.F.L. *An Economic History of New Zealand to 1939* Auckland (Collins) 1970 464p.
12. *New Zealand Herald* 4 June 1984 'Flaxwork grows into Art'.
13. *Rodney & Waitemata Times* 5 June 1984. Otamatea Section 'Gillian makes Baskets'.

14. Walls, G. *Traditional Uses of Plants in New Zealand and the Pacific* Summary Report (DSIR Botany Division) 1988 42p.
15. Collins D. *An Account of the English Colony of New South Wales* London 1798-1802 2 vols.
16. Thouin, A. (1803) 2 Annales du Muséum National d' Histoire Naturelle 228.
17. Saint Fond, Faujas de (1812) 19 Annales du Muséum d'Histoire Naturelle 419, pl. 20 *Phormium tenax*.
18. Moore, F., McNab, W.R. (1885) 'Reports for His Excellency the Lord Lieutenant on *Phormium tenax*', Botanic Garden, Glasnevin, 8p.
19. Norcliffe, G.B. (1969) 29(12) *J. Trop. Geogr.* 49.
20. Anon. (1966) 23(204) *Coopercotia* 38 (Commonwealth Agric. Bur. Hort. Plant Crops, Annotated Bibliog. 1974 New Zealand Flax 1961-1970 Abs. 5609).
21. Goulding, J.H. (1971) 8 *Rec. Auckland Inst. Mus.* 57.
22. *New Zealand Herald* 25 January 1989, Section 1, p. 9.
23. Monro, D. (1869, 2nd ed. 1875) 1 *Trans. N.Z. Inst.* 6.
24. Colenso, W. (1869, 2nd ed. 1875) 1 *Trans. N.Z. Inst.* 233.
25. Harris, W. (1984) 98 *Botany Division Newsletter* 3, 106 ibid. 4.
26. *New Zealand Herald* 21 January 1989.
27. Walls, G. (1990) 19 *DSIR Resource Research* 1.
28. Hamilton, A. *The Art Workmanship of the Maori Race in New Zealand* Wellington (New Zealand Institute) 1896-1900 439p.
29. Best, E. *Forest Lore of the Maori* Wellington (Government Printer) 1942 reprinted 1977, 420p.
30. Cunningham, A. (1836-1839) 2 *Comp. Bot. Mag.* 222, 327, 356. 1(3) *Ann. Nat. Hist.* 210, 1(5) ibid. 376, 1(6) ibid. 455, 2(7) ibid. 44, 2(8) ibid. 125, 2(9) ibid. 205, 2(11) ibid. 356, 3(14) ibid. 29, 3(15) ibid. 111, 3(17) ibid. 244, 3(18) ibid. 314, 4(21) ibid. 22, 4(22) ibid. 106, 4(24) ibid. 256. (The parts are usually bound in one volume as *Florae Insularum Novae Zelandiae Precursor, or a Specimen of the Botany of the Islands of New Zealand*).
31. Hunter, J. (1958) 2 *J. Roy. N.Z. Inst. Hort.* 342.
32. Anon. (1985) 100 *Botany Division Newsletter* 5, Harris, W. 105 ibid 3.
33. Goulding, J.H. (1964) *New Zealand Herald* 11 July, Section 3 'Brown Paper suitable for Banknotes'.
34. Murray, J. *An Account of the Phormium tenax or New Zealand Flax*. Printed on Paper made from its Leaves London (Henry Renshaw) 1836 34p. (Relfe & Fletcher) 1838 2nd ed. 56p.
35. New Zealand Exhibition 1865 *Reports and Awards of the Jurors* Dunedin (printed for the Commissioners by Mills, Dick & Co) 1866.
36. Kirk, T. (1871) 3 *Proc. N.Z. Inst.* 48.
37. Kirk, T. (1874) 6 *Trans. N.Z. Inst.* 55.
38. Angus, J.H. *Papermaking Pioneers* Mataura (N.Z. Paper Mills Ltd) 1976 212p.
39. Thornton, G.G. *New Zealand's Industrial Heritage* Wellington (Reed) 1982 194p.
40. Kirk, T. *The Forest Flora of New Zealand* Wellington (Government Printer) 1889 345p. 142pl.
41. Philpott, H.G. *A History of the New Zealand Dairy Industry 1840-1935* Wellington (Government Printer) 1937 413p.
42. Davis, M.D. *Making Paper in New Zealand* Nelson (Native Forests Action Council) 1982 18p.
43. Ushiyama, R., Bullivant, S., Matthews, R.E.F. 7 *N.Z.J. Bot.* 363.
44. *New Zealand Herald* 19 June 1986 'Flax proposed as Barrier on Motorways'.

7

Grasses

In pre-European times the Māori people had no grazing animals and no use for grass as pasture. Māori women wore sweet-scented kāretu (*Hierochloe redolens*) in waist girdles and necklaces. From Central Otago, Hamilton (1896) described a shoulder cape of flax to which small tussocks were tied, root end up, as an outer covering. Where nothing better was available, grass was used as thatch and floor-covering.[1]

Colenso (1872) wrote that the Māori recognized 'karetu', but lumped together all small grasses as 'patiti'; however, Allan in Volume 1 of the *Flora of New Zealand* (1961), and Williams in *A Dictionary of the Maori Language* (1971), identified a number of grasses under names such as 'pātītī', 'toetoe', and 'wī', with additional epithets to distinguish different kinds.[2-4]

When Captain Cook visited New Zealand in 1769-70, much of the country was covered in forest and scrub; grassland was restricted in the North Island to parts of the Volcanic Plateau, Hawke's Bay, and to the summits of the higher ranges between Wellington and East Cape. In the South Island, grassland formed a wider belt between the Southern Alps and the East Coast. During the following two centuries, much bush and scrub has been cleared and introduced grasses and clovers have been sown, garden plants have been established, and exotic weeds have flourished.

The indigenous grassland has been modified by burning, by the introduction of sheep, cattle, deer, rabbits, and other animals, and by the spread of exotic pasture plants and weeds and the application of fertilizers.

7.1 Beating about the Binomial Bush

Buchanan listed thirty-six Otago grasses, mostly natives, in his essay for the 1865 Exhibition, and classed twelve as 'first quality, forming the best pasture'; thirteen as 'second quality' and eleven as 'third quality'. His names for the twelve best grasses were: *Poa australis*, *P. breviglumis*, *Hierochloe redolens*,

Dichelachne crinita, Agrostis avenoides, A. quadriseta, A. parviflora, A. pilosa, A. aemula, Koeleria cristata, Trisetum antarcticum, and *T. subspicatum.* He considered that numerous Otago grasses were valuable for the fattening of stock, and they would probably repay the trouble and expense of being improved by cultivation, as they might prove more suitable to the climate and more nutritious than English grasses.

Buchanan deplored the repeated burning of the pasture in arid districts of Otago, pointing out that the finer grasses are destroyed by fire, or afterwards by sun or frost, while the coarser tussock grasses, spear-grass (*Aciphylla*), and many plants worthless as pasture, having large succulent roots, strike deep in the soil and are preserved. He considered that much of the grassland of Otago had been thus damaged, and that it was no wonder that many of the runs had deteriorated so that they required eight acres to feed one sheep.[5]

Kirk (1870) listed native and other grasses which were suitable for pastoral purposes in the Auckland province, restricting the native species to those decidedly preferred by cattle. He selected the best kinds of native and introduced grasses for a variety of habitats; for example, those for ordinary soils, swamps and meadows, clay soils, clay among mānuka, sand and muddy places near the sea, and so on. He urged settlers to study their land and use the grasses already present instead of spending money on grass seed from overseas which might not form permanent pasture.[6]

The Philosophical Institute of Canterbury formed a Committee to compile similar information about Canterbury grasses, and the Committee's report was published in 1872. Members of the Committee first listed thirty-three valuable native grasses, and asked selected farmers to complete a circular, but only three replied as most people approached were unable to identify their grasses according to the list of scientific names. The notes of the three farmers were supplemented with information from J.F. and J.B. Armstrong, two Christchurch botanists, and dealt with the palatability of each species. Kirk also sent the Committee notes, with the suggestion that the variety *cingula* of *Danthonia semiannularis* 'is a useful grass and appears to deserve the attention of the cultivator'.[7-11]

In 1880, Buchanan described all the known indigenous grasses. The illustrations are unusual in that they were made by inking dried grasses and placing them on the lithographic stone, details and enlargements being added by hand. Useful information on the 'forage potential' of each species was included in the text.[12]

In 1887, *A Manual of the Grasses and Forage Plants Useful to New Zealand,* by Thomas Mackay, was published by the Government Printer, Wellington. The Crown Lands Department outlined (but for some reason did not publish) Part 2, dealing with indigenous grasses, and Part 3, describing experimental work done on the cultivation and respective values of both

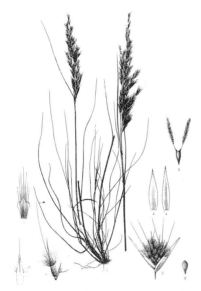

New Zealand oatgrass (*Rytidosperma* 'semiannularis',
formerly in *Danthonia* and *Notodanthonia*) **Poaceae**
Lithograph by John Buchanan (1880)

imported and native grasses for pasture and fodder. In 1936, H.H. Allan
produced *An Introduction to the Grasses of New Zealand*, briefly describing
and illustrating the introduced and more widespread native species.[13,14]

Formal taxonomic treatments of the indigenous grass flora were provided
by J.D. Hooker (1852-5, 1864-7), by J. Buchanan (1877-80), and by T.F.
Cheeseman (1906, 1925), and are now long out of date. However, another
volume of the *Flora of New Zealand*, prepared by Elizabeth Edgar and Henry
Connor, is nearing completion and will comprise keys, descriptions, and
illustrations of some 180 indigenous and 250 naturalized grasses. It will be the
first volume of the *Flora* in which both native and introduced species are
treated together.[12,15-18,53]

7.2 Transects and Quadrats

From A.D. Thomson's bibliography (1982), it is evident that Leonard
Cockayne was interested in the application of ecological methods to pasture
problems; his 'economic investigation' of montane tussock grassland, reported
in the *New Zealand Journal of Agriculture* between 1919 and 1922, involved a
study of the palatability for sheep of various pasture plants and forest forage
plants, the effect of spelling heavily grazed pasture, and the regrassing of

depleted areas in Central Otago. In 1926, a summary of these studies formed a chapter of Tansley and Chipp's classic work on *Aims and Methods in the Study of Vegetation.*[19-21]

An ecological study of Mount Peel was made by H.H. Allan in 1927, and a survey of all South Island grasslands was carried out by F.W. Hilgendorf in 1935. V.D. Zotov described South Island tussock grasslands in 1938, while A.P. Barker studied tussock grasslands at Hunters Hills, South Canterbury, in 1953. M.J. Wraight in 1963 described alpine and upper montane grasslands of the Wairau River catchment where spectacular erosion had occurred, while H. Connor in 1964 classified tussock grassland communities of the Mackenzie Country and in 1965 dealt with similar communities of the middle Rakaia Valley.[22-27]

From the launching of the *New Zealand Journal of Botany* in 1963, A.F. Mark contributed a series of papers on South Island vegetation, including the ecology of *Chionochloa rigida*, a dominant tussock on South Island high mountains. He recognized that some of the samples belonged to a hitherto undescribed species, now known as *C. macra*. Also from the inception of the new journal, H. Connor studied the morphology and breeding systems of native grasses while, from 1976, P.A. Williams and others assessed macro-element composition, growth, biomass, and net productivity of tussock grasses in both islands.[28-30]

7.3 Molesworth Station

The palmy days of tussock grassland were from 1870 to 1880, when sheep numbers grew from one million to ten million. In the same decade, rabbits, which had been introduced with little apparent success from the 1830s on, increased to vast numbers in many South Island localities (severe infestations in parts of the North Island were reported in the 1890s). In 1867, Lady Barker wrote of the fun of 'burning the run', and annual fires were a part of sheep farming at that time. So while sheep numbers increased dramatically, a combination of burning-off, rabbits, and sometimes overstocking, had disastrous effects on the plant cover on many sheep runs.[31]

A.H. Cockayne, son of the illustrious Leonard Cockayne, warned in the early 1900s that the recovery of damaged tussock grassland would be extremely slow; Lucy Moore's account of the changing vegetation of Molesworth Station between 1944 and 1971, published as a DSIR Bulletin in 1976, recorded slow recovery such as that foreseen by Cockayne.[32,33]

Molesworth farm, which covers 182,000 ha and is roughly 64 km by 64 km, supported 97,000 sheep in 1900, but repeated burning, rabbits, and overstocking resulted in severe deterioration of the pastures, and the Government resumed control. Stocking was reduced to about 9500 cattle, 300 sheep, and 75 horses. On suitable flats, oversown with introduced grasses and clovers, pastures grazed by cattle have been maintained. On the hills once covered in tussock there has been little recovery or spread of native species despite the removal of sheep, cessation of fires, reduction in the number of rabbits, and oversowing by air with introduced plants which have the ability to persist. Lucy Moore found that any recovery on bare ground was due initially to introduced weeds such as sorrel, mouse-ear hawkweed, Yorkshire fog, and so on.

In 1987, the Deputy Prime Minister, Hon. Geoffrey Palmer, ruled that Molesworth would remain in Crown ownership with Landcorp managing the farming and working with the Department of Conservation on a management plan to safeguard scenery and conservation values and facilitate public access. Difficulties subsequently arose regarding the management plan and public access to Molesworth, and the Royal New Zealand Forest and Bird Protection Society called on the Conservation Minister, Helen Clark, to make it clear that cattle-grazing was not the sole purpose of the property.[34]

7.4 Reserving Remnants

In the North Island, much of the tussock shrubland is situated in the Tongariro National Park and in forest parks such as the Kaimanawa Forest Park, and is protected. Ling (*Calluna vulgaris*), purposely sown by a ranger, has spread widely in grassland and open shrubland at Tongariro, and seedling pines from plantations upwind have infested the south-west corner of the Park.

In the South Island, about half of the original tussock grassland has gone. G. McSweeney and L. Molloy, in the November 1984 issue of *Forest & Bird*, estimated that areas totalling about 1.5 million ha of tall tussock (*Chionochloa* species) have survived at higher altitudes, but only 650,000 ha of short tussock (*Festuca* and *Poa* species) remain at lower altitudes, much having been destroyed by overgrazing or replaced by introduced grasses. In the intermediate montane zone, snow tussock (*Chionochloa* species) remains on about 600,000 ha, but is under attack as intensive agriculture replaces traditional sheep farming, as further areas are irrigated from hydro-electric schemes, and as exotic forests are planted. Southland red tussock grassland has been replaced almost entirely by exotic pasture, less than 10 ha of red tussock being reserved on the entire Southland Plain.[35]

Reserve needs were defined by the Protected Natural Areas Programme, and on the publication in 1984 of the *Register of Protected Natural Areas in New Zealand*, the dearth of reserves in the central and eastern South Island, especially on Crown pastoral leasehold land, became immediately apparent. Not only were tussock grassland reserves inadequate, but wetland, shrubland, and duneland protected areas were sadly lacking.[36]

In recent years the tourist industry, with ever-increasing numbers of visitors to cater for, has sought to have the natural areas along scenic routes preserved intact; the National Parks and Reserves Authority has pressed for a system of reserves fully representative of natural landscapes; the National Soil and Water Conservation Authority has stressed the need for protection of natural areas to check water run-off and soil erosion, and the Forest and Bird Protection Society, Federated Mountain Clubs, Deerstalkers, Acclimatization Societies, and others concerned with the outdoors, have joined together to seek more reserves for their special interests. During 1984, six brief articles identifying specific areas of tussock grassland which should be preserved were published in *Forest & Bird*.[37-43]

Despite all this concern, progress was slow. Although most South Island tussock grassland is owned by the Crown, a great deal of it is held by leaseholders who have perpetual right of renewal, and are concerned mainly with freeholding their runs. In 1984 the Labour Party, in its election manifesto, stated that it would allow freeholding only when there was a 'comprehensive protection of soil and water values and recreational environmental values'. The Labour Government has subsequently revamped Lands, Forestry, and Conservation administration, and some of the reserves sought will probably be attained. Already some progress has been made; for example, in Otago, an area of 400 ha has been set aside as the Lindis Pass Scenic Reserve, while in Southland 3100 ha of red tussock and beech at Gorge Hill alongside the highway to Te Anau have been protected for a takahē habitat and breeding programme.

In 1986, the Government decided that ownership of pastoral leases in the South Island high country should remain with the Crown. Deputy Prime Minister Geoffrey Palmer and the Lands and Conservation Ministers Koro Wetere and Russell Marshall issued a joint statement that Cabinet had decided that the protection of the lessees' rights and the rights of the interests of the public were of paramount importance. Landcorp was given the task of administering the leases. The Department of Conservation was charged with the protection of the native plants and animals and the unique high country landscape, soil and water conservation, and the negotiation of increased opportunities for public recreation.

Conservationists and members of Acclimatization Societies, Mountain and Tramping Clubs, and other people with an interest in the 'outdoors' welcomed the Cabinet decision. However, high country pastoral leases have farming as well as conservation priorities, and conflict seems inevitable. The Public Lands Coalition, formed by Acclimatization Societies, Federated Mountain Clubs, and the Forest and Bird Protection Society, is pressing for the retention of Crown ownership of all lands that are of value for conservation and recreation, protection of native plants and animals, public access rights to conservation lands, and so on. However, the picture is confused by the Ngai Tahu Māori land claim to all the land in the high country pastoral leases. The claim is before the Waitangi Tribunal, and the Public Lands Coalition has advocated that, whatever the Tribunal may recommend, protection of the 'public interest' should be paramount in any settlement of the Ngai Tahu claim.[44-46]

7.5 Better and Better Pastures

The story of A.H. Cockayne is very much the story of pasture farming in New Zealand from the end of the nineteenth century to the late 1930s. Bruce Levy, who served for a while as A.H. Cockayne's assistant, recorded both stories in *Grasslands of New Zealand* (3rd edition 1970). A.H. Cockayne advocated the establishment of a station to improve pasture plants, as well as soils and farm animals. He considered that plant breeding was necessary to obtain better seed, and suggested seed testing and seed certification twenty years or more before these facilities were accepted as necessary. A.H. Cockayne and Bruce Levy began selecting better grass strains as early as 1910. They worked in the Biological Laboratory of the Department of Agriculture, from which evolved a number of scientific institutions such as Grasslands Division, Botany Division, and Plant Diseases Division of DSIR.

The early breeding programme of Grasslands Division was criticized for not attempting to improve native grasses (such as the *Danthonias* for hard conditions and low-fertility soils), but Levy explained that improvement in the productivity of pasture was the main objective. Better soil fertility had to be attained before higher productivity was possible; once soil fertility was reached by top-dressing, it was better to use higher-yielding species already available. Consequently, ryegrass and clover were sown.[32]

Today, most New Zealand pastures are planted in ryegrass–clover strains that Grasslands Division has developed and tested. The Division is based at Palmerston North, with field stations in different regions. It produces approximately two new strains of pasture plants each year. From time to time, pastures become infected with new fungal and viral diseases, and indigenous grasses may eventually have to be considered for plant breeding; for this reason, if for no other, native grasses should be protected and perpetuated.

7.6 Sports Turf

Smaller plants are preferred for bowling greens and tees, hardier plants for the rough conditions of playing fields, and larger plants are acceptable for fairways, aerodromes, and racetracks. This is a grouping of turf plants indicated by G.S. Harris in *Turf Culture* (1972), prepared by the N.Z. Institute of Turf Culture. The turf plants described by Harris fall into two main groups:

some are grasses, and a few are 'weed' species of dicotyledonous genera such as *Cotula, Plantago, Dichondra*, and *Hydrocotyle*. Both groups of plants are of prostrate, spreading habit, and produce a close turf under regular mowing.[47]

Among the grasses described by Harris were six native kinds. *Festuca rubra* ssp. *commutata*, 'first found in the South Island of New Zealand where it grows abundantly on the lighter and poorer soils', is 'one of the best turf-formers, ideal for putting greens'. It is 'easily mown, can be cut regularly and closely, withstanding heavy dressings of . . . fertiliser . . . sprays, and remains green summer and winter'. It 'forms a better turf when combined with browntop than when growing alone' and is 'harvested for seed' in the South Island.

Microlaena stipoides was described as a 'good turf-former where it is already established on fairways. Withstands close mowing'. For some reason, 'seed is not available commercially'. *Danthonia pilosa*, identified as one of the *Notodanthonia biannularis* group, was rated less highly: 'makes a dense mat of wear-resisting turf for fairways' and 'is encouraged by close mowing which it withstands well', but 'it is extremely slow to establish from seed', 'it will not tolerate fertile soils and . . . fertilisers', and 'wherever it will grow well, Chewings fescue should do equally well or better'.

Danthonia semiannularis, identified as another of the *Notodanthonia biannularis* group, was regarded by Harris as inferior to *D. pilosa*, while *Deyeuxia billardieri* and *Lachnagrostis filiformis* (formerly *D. filiformis*), although included as turf plants, were both described as 'no use in turf', probably because they are annuals. Species of *Notodanthonia* Zotov have been transferred to *Rytidosperma* Steudel, an earlier name.[54]

It is puzzling that none of the small, cushion-forming alpine grasses are suitable for turf, especially in the southern part of New Zealand. Possibly their role has yet to be determined. The New Zealand species of *Zoysia* might yet be found suitable for turf — one species used to form patches of dense cover on consolidated sand dune at Piha, before lupins became dominant.

7.7 Future Directions

In 1983, the N.Z. Forest Service reviewed all major protection forest rehabilitation schemes and found that an area of about 500,000 ha of South Island high country would benefit from 'artificial revegetation', while 600,000 ha of North Island country could benefit from such treatment, given intensive animal and fire control.[49] Grasslands Division, the National Plant Materials Centre of the Ministry of Works and Development, the Protection Forestry Division of the Forest Service, and the National Soil and Water Conservation Authority all seem to be concerned with grasses and other plants for erosion control and the revegetation of eroded surfaces, and in the past have advocated both introduced and native species. Much of the land in question is now under the administration of the new Department of Conservation, or subject to various Māori claims, and the future of the scheme is unclear.

In the Annual Report of DSIR for 1986, mention was made of the massive reduction in top-dressing on hill country, the severe risk of the country reverting to scrub and weeds, and the possibility of using goats as a cheaper, non-herbicidal method of weed control. Grasslands Division was evaluating growth potential, nutritive value, and palatability to sheep and goats of such common pasture weeds as mānuka and toetoe. The Report predicted that 'forest grazing' would undergo a huge expansion in the next twenty years, and that research involving a number of disciplines would be necessary into 'the complex-interactions between different animal species, grassland, browse shrubs and forestry plantations'.[50] Under the new system of 'user pays', the future of this scheme is also unclear.

Two recent studies by staff of Botany Division, DSIR, indicate that better management of indigenous tussock grassland may be possible and worthwhile.

R.B. Allen and T.R. Partridge (1988) studied the effects of fire on two adjacent areas of snow tussock (*Chionochloa rigida*) at about 560 m altitude on Flagstaff, near Dunedin. One area was burned after the dry summer and autumn of 1976, the other in the moister conditions of the following spring. The severe autumn fire destroyed the litter layer, and only plants able to resprout from roots or buried stems recovered. Other species had to re-establish from seeds, or disappeared from the burnt area. The spring fire did not destroy the litter layer completely, and prostrate plants were able to recover more quickly. After nine years there were still some differences in plant cover between the autumn and spring fire sites. In both areas, the increase in the size of snow tussocks and mountain flax plants (*Phormium cookianum*) resulted in the suppression of small grasses, herbs, and shrubs that initially occupied the spaces opened up by the fires.[51]

In the snow tussock grassland of Otago, several exotic conifers have become established as weeds of pasture and conservation areas. Ralph Allen and Bill Lee (1989) studied conifer establishment at three sites with relatively intact snow tussock (*Chionochloa rigida*) grassland: corsican pine (*Pinus nigra*) at Naseby, lodgepole pine (*Pinus contorta*) at Coronet Peak, and larch (*Larix decidua*) at Mt Aurum. They found that conifer seedlings established most frequently on the southern (moister) side of tussocks, especially where the ground was bare or nearly so. Larch seedlings were most frequently associated with the shorter alpine fesce tussock (*Festuca matthewsii*). Perhaps the deciduous habit of larch reduced the need for winter shelter. The authors suggested that development of continuous tussock canopy and elimination of exotic conifer seed sources as part of grassland management would reduce the weed problem.[52]

1. Hamilton, A. *The Art Workmanship of the Maori Race in New Zealand* Wellington (N.Z. Institute) 1896-1900 439p.
2. Colenso, W. *Ko te A-nui A Wi, nei ako maana ki te reo Ingarahi: Willie's First English Book* Wellington (Government Printer) 1872 2 vols.
3. Allan, H.H. *Flora of New Zealand, Volume 1* Wellington (Government Printer) 1961 1085p.
4. Williams, H.W. *A Dictionary of the Maori Language* Wellington (Government Printer) 1971 7th ed. 507p.
5. Buchanan, J. (1869, 2nd ed. 1875) 1 *Trans. N.Z. Inst.* 181.
6. Kirk, T. (1870) 2 *Trans. N.Z. Inst.* 102.
7. Canterbury Philosophical Institute (1872) 'Report of a Committee on Native and Introduced Grasses, with Appendices', 4 *Trans. N.Z. Inst.* 292.
8. Boys, J.C., Dixon, M. Glyde, S.D. (1872) 4 *Trans. N.Z. Inst.* 304.
9. Armstrong, J.F. (1872) 4 *Trans. N.Z. Inst.* 309.
10. Armstrong, J.F., Armstrong, J.B. (1872) 4 *Trans. N.Z. Inst.* 301.
11. Kirk, T. (1872b) 4 *Trans. N.Z. Inst.* 295.
12. Buchanan, J. *The Indigenous Grasses of New Zealand* Wellington (Government Printer) 1877, 1879, 1880, 3 parts.
13. Mackay, T. *A Manual of the Grasses and Forage Plants of New Zealand.* Wellington (Government Printer) 1887 164p. 47pl.
14. Allan, H.H. *An Introduction to the Grasses of New Zealand* Wellington (Government Printer) 1936. DSIR Bull 49 159p.
15. Hooker, J.D. *The Botany of the Antarctic Voyage of H.M. Discovery Ships Erebus and Terror in the years 1839-1843* London (Reeve Bros) 1844-1860 6 parts in 3 vols.
   Vol.1, 2 parts: *Flora Antarctica* 1844, 1847.
   Vol.2, 2 parts: *Flora Novae Zelandiae* 1852, 1855.
   Vol.3, 2 parts: *Flora Tasmaniae* 1855, 1860.
16. Hooker, J.D. *Handbook of the New Zealand Flora* London (Lovell Reeve) 1864-1867. Vol. 1, 1864, 392p; vol. 2, 1867, p. 393-798.
17. Cheeseman, T.F. *Manual of the New Zealand Flora* Wellington (Government Printer) 1906 1199p.
18. Cheeseman, T.F. *Manual of the New Zealand Flora* Wellington (Government Printer) 1925. 2nd ed., revised by the author, edited by W.R.B. Oliver, 1163p.
19. Thomson, A.D. (1982) 20 *N.Z.J. Bot.* 205.
20. Cockayne, L. (1919) 18 *N.Z.J. Agric.* 1, 321; 19 ibid. 29, 129, 343; (1920) 20 ibid. 82, 209, 337; 21 ibid. 176, 324; (1921) 22 ibid. 148; 23 ibid. 137; (1922) 24 ibid. 321, 25 ibid. 1, 129.
21. Tansley, A.G., Chipp, T.F. (eds) *Aims and Methods in the Study of Vegetation* London (British Emp. Veg. Committee) 1926, pp. 274, 330, 349, 362.
22. Allan, H.H. (1927) 57 *Trans. N.Z. Inst.* 73.
23. Hilgendorf, F.W. *The Grasslands of the South Island of New Zealand* Wellington (Government Printer) 1935; DSIR Bull. 47, 24p; Zotov, V.D. (1938) 20A *N.Z.J. Sci. Tech.* 212, (1939); DSIR Bull. 73, reprinted 1947.
24. Barker, A.P. *An Ecological Study of Tussock Grassland, Hunters Hills, South Canterbury* Wellington, (Government Printer) 1953; DSIR Bull. 107, 58p.
25. Wraight, M.J. (1963) 1 *N.Z.J. Bot.* 351.
26. Connor, H.E. (1964) 2 *N.Z.J. Bot.* 325.
27. Connor, H.E. (1965) 3 *N.Z.J. Bot.* 261.
28. Connor, H.E., Matthews, B.A. (1977) 15 *N.Z.J. Bot.* 531 and earlier papers.

29. Mark, A.F. (1965) 3 *N.Z.J. Bot.* 73, 180, 277, 300.
30. Williams, P.A., Mugambi, S., Nes, P., O'Connor, K.F. (1978) 16 *N.Z.J. Bot.* 479 and earlier papers.
31. Barker, Lady *Station Life in New Zealand* London (MacMillan) 1867. 237p. Reprinted by Whitcombe & Tombs 1950, 1951.
32. Levy, E.B. *Grasslands of New Zealand* Wellington (Government Printer) 1970, 3rd ed., 374p.
33. Moore, L.B. *The Changing Vegetation of Molesworth Station, New Zealand, 1944 to 1971* Wellington (DSIR Science Information Division) 1976; DSIR Bull. 217 118p.
34. (1987) 40 *Forest & Bird Conservation News* 3.
35. McSweeney, G., Molloy, L. (1984) 234 *Forest & Bird* 2.
36. Department of Lands & Survey *Register of Protected Natural Areas in New Zealand* Wellington (Government Printer) 1984, 468p.
37. McSweeney, G. (1983) 229 *Forest & Bird* 50.
38. Hutchins, L. (1984) 234 *Forest & Bird* 6.
39. Patrick, B. (1984) 234 *Forest & Bird* 7.
40. Mark, A. (1984) 234 *Forest & Bird* 8.
41. Mason, B. (1984) 234 *Forest & Bird* 9.
42. Dunnett, B. (1984) 234 *Forest & Bird* 11.
43. Molloy, L. (1984) 234 *Forest & Bird* 12.
44. Public Lands Coalition (1988) *Public Lands News* November, 4p.
45. (1989) 50 *Forest & Bird Conservation News* 5.
46. (1990) 60 *Forest & Bird Conservation News* 3.
47. Harris, G.S. 'Constituents of Turf' in N.Z. Institute of Turf Culture *Turf Culture* Palmerston North (Simon Printing) 1972 362p, p. 25.
49. New Zealand Forest Service *Report of the Director-General of Forests for the year ended 31 March 1983* p. 14.
50. New Zealand Department of Scientific and Industrial Research *Report for the year ended 31 March 1986* pp. 13, 33.
51. Allen, R.B., Partridge, T.R. (1988) 122 *Botany Division Newsletter* 8.
52. Allen, R., Lee, B. (1989) 27 *N.Z.J. Bot.* 491.
53. Webb, C.J. (1989) 17 *N.Z. Bot. Soc. Newsletter* 8.
54. Connor, H. E., Edgar, E. (1979) 17 *N.Z.J.Bot.* 311.

8

Food and Beverages

8.1 Root Crops

CABBAGE TREES

In pre-European New Zealand, kūmara (*Ipomoea batatas*), taro (*Colocasia esculenta*), yam (*Dioscorea alata*) and tī (*Cordyline terminalis*) were cultivated as root crops. All were brought to New Zealand from the tropical Pacific by the Polynesian ancestors of the Māori people.[1]

Barry Fankhauser, Chemistry Department, Chaminade University of Honolulu, Hawaii, studied New Zealand cabbage trees about 1985 while he was a Ph.D. candidate in the Anthropology Department of the University of Otago, Dunedin. In his thesis and in several papers he discussed the botany of *Cordyline* (complicated by hybrids), the ethnobotany, native names, cultivation, storage and cooking of the plants.[2-4]

For food, the Māori used four endemic species of cabbage tree (*Cordyline australis, C. banksii, C. indivisa,* and *C. pumilio*) and one introduced species (*C. terminalis*). *Cordyline pumilio* and *C. terminalis* have only an edible 'root' (underground stem), while the other three species also have an edible aerial stem and top (consisting of the leaf bases).

In Taranaki, a tī known as 'tawhiti' was planted over large stones so that the root did not grow too large, thereby making it easy to dig up. A Māori possessing a 'rua tī' or tī store house was considered to be a person of standing.

Cordyline australis is the commonest cabbage tree in New Zealand and was the most widely eaten species in the South Island, where large pits still indicate the scale on which the Māori grew and cooked the plant. It will grow almost anywhere, and it responds to good soils and conditions. It varies in composition according to season and has the best food value in the late spring and early summer. The texture of tī is fibrous, with starch and sugars interspersed among the fibres, as with sago. The fibre could be separated by long cooking or by breaking up the root manually beforehand. Hydrolysis of

Table 8.1

FOOD COMPOSITION, NUTRITIVE VALUE (%), EDIBLE PORTION

Food	Water	Ash	Fat	Carbo-hydrate	Fibre	Protein	Kcal 100g	kJ 100g
Potato	79.8	0.9	0.1	17.1	0.5	2.1	76	318
Sweet Potato	70.6	2.0	0.4	26.3	0.7	1.7	114	477
Taro	72.5	1.2	0.2	24.2	0.9	1.9	104	435
Yam	73.5	1.0	0.2	23.2	0.9	2.1	101	423
Wheat-cooked	87.7	0.8	0.3	9.4	0.3	1.8	45	188
Cordyline australis:								
Root	64.0	0.9	1.4	23.6	10.3	0.4	103	431
Stem	68.3	0.7	1.5	14.9	13.6	0.4	71	297
Top	81.5	1.7	3.2	8.8	4.5	1.4	68	283

the polysaccharides present in the root gives over 90% fructose, which is much sweeter than any other common sugar.

Fankhauser's thesis included more exhaustive analyses than are available for any other native food plant; Table 8.1 contains data from the thesis. The lipids were analysed by M.H.G. Munro.[5] Fankhauser concluded that the root of *C. australis* has too little protein to be an adequate source, but as it is usually eaten with high protein animal foods, this is not of great significance. It will be noted from the table that the root has a rather high fat content compared with other roots.

Fankhauser and Brasch (1985) suggested that a new industry to produce fructose by hydrolysis of cabbage tree 'roots' might be viable.[4] Jay Mann (Biotechnology Division) and Katrina Stewart (Botany Division DSIR) reported on a trial planting of *C. australis* at Lincoln where, after a year in the field, the plants were harvested, dried, and analysed. The plants yielded about four tonnes of fructose per hectare. This was valued at $12,000, which suggested that further work might be worthwhile.[6]

Since 1987, cabbage trees in the northern part of the North Island have been affected by a new disease, so far known only as the 'Rapid Decline Syndrome'. Many trees are affected, and until the cause is determined and a remedy is found, prospects for a *Cordyline australis* fructose industry are clouded.[7,8,61]

BRACKEN
The root (rhizome) of bracken or aruhe (*Pteridium esculentum*) was a major vegetable food of the Māori in many parts of New Zealand. Varieties of the plant grow wild in many parts of the world and it is not considered much of a foodstuff. A.M. Smith and E.W. Fenton, who carried out an investigation of

Scottish bracken root during the Second World War, concluded (1944) that it would be better to let the roots rot and use them for manure.[9]

Before eating the roots, the Māori soaked, roasted, and pounded them. They were often flavoured with cabbage tree fructose, flax nectar, or honey. The starch was also separated from the fibre and cooked in a cake (komeke), sometimes flavoured with the juice of tutu berries (*Coriaria* species). The Māori people knew of areas where large bracken roots grew, and occasionally burnt the fronds at the right time of the year to stimulate fresh growth.

Andrew Crowe (1981) recorded the reactions of people who have eaten fern root. Dr L.K. Gluckman (1976) quoted a number of reports on the extremely constipating effect of the root and the explorer Thomas Brunner found that at times he suffered intense pain from eating this food. D. Burton (1982) wrote that the fibres were normally removed beforehand, which may reduce the problem.[10-12]

P. Houghton (1980) estimated that, in prehistoric times, the average age of the Māori at death was thirty-one or thirty-two years. The possible carcinogenic effects of heavy eating of fern root may have been a contributing factor (Nicholls); however, the average age of death in prehistoric societies in America, Europe, and Asia was also in the thirties.[13,14]

Other roots that were eaten raw or roasted in former times included taramea or spaniard (*Aciphylla colensoi* and *A. squarrosa*), rengarenga (*Arthropodium cirratum*), rīrīwaka (*Bolboschoenus caldwellii, B. fluviatilis* and *B. medianus*), pōhue (*Calystegia sepium*), shore bindweed (*C. soldanella*), perei or potato orchids (*Gastrodia cunninghamii* and *G. sesamoides*), cranesbill (*Geranium homeanum, G. solanderi,* and allied species), native scorzonera (*Microseris scapigera*), onion orchid (*Microtis unifolia*), māikaika (*Orthoceras strictum, Thelymitra pulchella*), maikuku (*Thelymitra longifolia*), and raupō (*Typha orientalis*).[15,16] Most of these herbs are rare and local, or becoming so, and they should not be gathered for food except in an emergency, or from plants grown for the purpose. The carrot-like root of the taramea was difficult to gather as the plant is spiny and some people used a rope to remove the top.

8.2 Leafy Vegetables

FIDDLEHEADS

Kiokio (*Blechnum* species) is reported to be a food plant throughout Polynesia, and there could be room for more investigation of fern fronds as green vegetables. Most recent research has been concerned with the toxic and carcinogenic nature of the foliage. Crowe (1981) listed as edible the young fronds or fiddleheads of several species including hen and chickens fern (*Asplenium bulbiferum*), shining spleenwort (*A. oblongifolium*), hard ferns

(*Blechnum* species), parsley fern (*Botrychium australe*), common shield fern (*Polystichum richardii*), and bracken (*Pteridium esculentum*). Parsley fern (*Botrychium australe*) is rare and local however, and should not be disturbed. Crowe described *Blechnum* fiddleheads as 'delicious boiled for fifteen minutes and dressed with oil, lemon juice, garlic and salt'.[10]

ANTISCORBUTICS
In Chapter 2 we mentioned Captain Cook's successful use of greens in salads and boiled for the treatment of scurvy. Among the New Zealand plants tried were: 'Cabbage tree', 'celery', 'spinach', 'sow thistle', 'scurvy grass' and 'a herb like that which Countrey people in England call Lambs Quarters or Fat Hen'.
   At Tolaga Bay in October 1769, a single 'Cabbage tree' was cut down and the cabbage used for a 'delicious meal'. In his journal, Banks referred to it as a 'palm cabbage' and J.D. Hooker, who edited the journal in 1896, identified it as *Areca sapida*, now known as nīkau (*Rhopalostylis sapida*).[17]
   At Dusky Sound, between March and May 1773, the hearts of a number of 'cabbage trees' were cut for food. George Forster wrote that, when prepared with oil and vinegar, these rolled up leaves were among the best salads he had eaten.[18] He identified the 'cabbage tree' as *Dracaena indivisa*, now known as tōī (*Cordyline indivisa*). The heart of the common cabbage tree (*C. australis*) was described as edible by Taylor (1855), Colenso (1869), and Crowe (1981), although all agreed that it is bitter raw and improved with cooking.[10,16,19]
   'Celery' has been identified as *Apium australe* (larger plants) and *A. filiforme* (smaller plants), and both are now known as *A. prostratum*. French explorers collected seed of the plant in Tasmania and it was grown and flowered at Malmaison, the garden of the Empress Josephine. An etching of the plant by Pierre-Joseph Redouté was published in 1804.[20]
   'Spinach' was among the seeds that Banks and Solander took back to England in 1771. In Chapter 3 we described how the plant was grown in Moscow and named at Leningrad in 1781 and at Göttingen in 1785. This was the beginning of a very extensive literature on the cultivation and properties of New Zealand spinach (*Tetragonia tetragonioides*), which persists as a minor leafy vegetable in British, European, and American gardens.
   'Sow thistle' was described by Banks as 'exactly the same as in England'. George Forster (1786) identified it as *Sonchus oleraceus*, and commented that 'its delicate stems and young leaves made a salad for our enjoyment'. The identity of 'sow thistle' is discussed in Brooker, Cambie and Cooper (1987). The latest information (Webb, Sykes, and Garnock-Jones, 1988) is that early records, in which sow thistle was identified as *S. asper*, related to *S. kirkii*.[17,18,21-23]
   The names 'pūwhā' (or 'pūhā') and 'rauriki' are applied by the Māori to sow thistles which are eaten. 'Pūhā pororua' is preferred because of its soft, smooth

leaves. Colenso (1881) and Crowe describe how sow thistle stems are bruised and the bitter milky juice is washed away.[10,15]

In 1983, the Marlborough Development Council initiated a project to grow pūwhā commercially, and an application was made for government assistance. Pūwhā is available in the markets in Auckland and Wellington. Sow thistle is widely used as a vegetable overseas, and Burton has modern recipes containing pūwhā as an ingredient.[12]

Cook's 'scurvy grass' (*Lepidium oleraceum*) was plentiful on the rocky coast of New Zealand, and Cook's men gathered it by the boat-load. It is found now only on islets inaccessible to goats and farm stock. The white butterfly may have helped its destruction. *Lepidium oleraceum* has a number of forms and several allied species, all now rare. The complex might provide new and valuable green vegetables if collected, grown and studied. Moore and Irwin (1978) recommended small quantities of the leaves in salad and their sharp cress taste might be acceptable in sandwiches.[23-25]

'A herb like . . . Lambs Quarters or Fat Hen' has been identified as *Tetragonia* (J. Anderson 1821), *Atriplex patula* (J.D. Hooker 1896) and *Chenopodium album* (Morrell 1958). *Tetragonia* is New Zealand spinach. *Atriplex patula* and allied species and *Chenopodium album* are regarded as post-European introductions to the New Zealand flora in the latest treatment (Webb, Sykes, and Garnock-Jones).[17,23,26,27]

OTHER NATIVE GREENS

Andrew Crowe discussed other native herbs which may be served as cooked green vegetables including hawkweed (*Picris hieracioides*), panakenake (*Pratia angulata*), poniu (*Rorippa palustris*), glasswort (*Sarcocornia quinqueflora*), pūnui (*Stilbocarpa polaris*), native dandelion (*Taraxacum magellanicum*), and dwarf bush nettle (*Urtica incisa*).[10]

Pūnui (*Stilbocarpa polaris*) was known to sealers and whalers as Macquarie Island cabbage and eaten as a remedy for scurvy. Found only on the Auckland, Campbell, and Macquarie Islands, it has been reduced by sheep and rats. Bellinghausen, a Russian explorer who visited Macquarie Island in 1820, described soup and pickles made from the stalks and roots, and Baron von Mueller of Melbourne recommended the plant as a vegetable for colder climates. B.W. Taylor (1955), quoted in Crowe, described the cooked leaf stalks as like celery, the pickled rhizomes as like turnips and the cooked leaves as like wet blotting paper.[10,28,29]

Iti lacustris is a small native cress found on pebble beaches of lakes Manapouri and Te Anau. When P.J. Garnock-Jones and P. Johnson named the plant in 1987, they invited readers to write to Botany Division, DSIR, for seed and to help conserve the species. The plant is now in cultivation and is grown by enthusiasts as an edible cress.[30]

8.3 Fruits

GYMNOSPERMS
Native gymnosperms with edible fruits include kahikatea (*Dacrycarpus dacrydioides*), rimu (*Dacrydium cupressinum*), tōtara (*Podocarpus totara*), miro (*Prumnopitys ferruginea*) (Plate 5) and matai (*P. taxifolia*). Kahikatea, rimu and tōtara have an edible fleshy receptacle beneath the ripe seed. Miro and matai have drupelike fruits in which a large, hard seed is surrounded by a thin layer of edible flesh. All were gathered and eaten by the Māori, although tōtara and miro fruits have a turpentine flavour. Crayon (1842), a writer in the *New Zealand Journal*, reported that at one Māori feast there were sixty baskets of 'kaikatea berries'.[31]

MONOCOTYLEDONS
Native monocotyledons with edible fruits include kōwharawhara (*Astelia* species), kahakaha (*Collospermum* species), kiekie (*Freycinetia baueriana* ssp. *banksii*, known here as *F. banksii*), and supplejack (*Ripogonum scandens*) (Plate 8). Crowe (1981) found berries of kōwharawhara (*Astelia banksii*) too bitter and preferred the base of the flower stalk.[10] The leaves or bracts of the spadix of kiekie have swollen edible bases, described as 'delicious'; however, those who have tried the centre of a female phalange are less enthusiastic. The red flesh of supplejack berries is described as rather tasteless and dry. The best food-plant of the group is kiekie, and the current programme of the Forest Research Institute and Department of Conservation to ensure supplies of kiekie leaves for weaving might well be widened to include plants with larger edible flower bracts.

DICOTYLEDONS
Edible fruits of native dicotyledons include berries (pulpy fruit containing seeds), drupes (hard stones inside a fleshy covering), aggregated drupelets (like a blackberry in form), and various swollen succulent petals, tepals, and arils.

BERRIES
Some berry fruits are toropapa (*Alseuosmia macrophylla*), wineberry (*Aristotelia serrata*), kōtukutuku (*Fuchsia excorticata*) (Plate 9), ramarama (*Lophomyrtus bullata*), autetauranga (*Pimelea arenaria*), pinātoro (*P. prostrata*), tawāpou (*Planchonella costata*), poroporo (*Solanum aviculare* and *S. laciniatum*), mairetawhake (*Syzygium maire*), and kōhia (*Tetrapathaea tetrandra*) (Plate 12).
    Wineberry fruit have been used for jams, jellies, and wine. Kōtukutuku fruit are sweet, though rather astringent, and the Reverend Vicesimus Lush described a pudding made from the berries, which met with a mixed

reception.[32] The flesh of tawāpou berries was eaten by northern Māori. It is one of the most handsome berries in the native flora, and the tree should be planted to conserve it. The hard seeds were used by the Māori for necklaces and are still in demand for the purpose. Ripe poroporo berries have been used for jam, although the green fruits may be poisonous. Both *Solanum* species are the source of material overseas for the synthesis of steroidal drugs.

DRUPES

Some edible drupes are taraire (*Beilschmiedia tarairi*), tawa (*B. tawa*), karaka (*Corynocarpus laevigatus*), hīnau (*Elaeocarpus dentatus*), kawakawa (*Macropiper excelsum*), and maire (*Nestegis cunninghamii* and *N. lanceolata*). Both taraire and tawa have drupes with edible flesh and kernels. They are discussed at some length by Crowe, who found the flesh acceptable, and the kernels edible after roasting or boiling; they kept at this stage. Best (1942) recorded that the kernels of taraire and tawa were highly prized by the Māori and were the staple food of the Tuhoe people of the Urewera.[10,33]

Karaka is mentioned in Chapter 9 (Potions and Poisons) and in Chapter 10 (Chemical Products from Native Plants). Illness resulting from eating karaka berries was described by Gluckman as 'the second indigenous disease' and his book has a paragraph on the symptoms and treatment. Opinions differ on how the drupes should be treated to remove the poisonous glucoside, karakin; Burton suggested three months immersion in water after the fruits have been steamed for twenty-four hours.[11,12] At Botany Division, DSIR, Lincoln, Brian Molloy (1990) grew five species of *Corynocarpus* and compared the four Australian and Melanesian species with the indigenous *C. laevigatus*. All have larger fruits and are more frost-tender than karaka. He noted that karaka is a reliable cropper and there is scope for the selection of trees or for a breeding programme to obtain larger fruits, and for an investigation of the processing and nature of the food product.[34]

Hīnau drupes were soaked in water, rubbed between the hands, and the flesh was baked into a cake with an oily flavour, or made into a gruel that was said to be good for invalids. 'If you wake me from my sleep, let it be for hīnau bread' was a Māori proverb. Crowe (1981) emphasized that only hīnau flesh was used; the hard kernels were thrown away.[10]

DRUPELETS AND AGGREGATED DRUPELETS

Karamū (*Coprosma lucida*) drupelets were listed as edible by J.D. Hooker (1864) and the fruits of this and allied species are still eaten by children as a bush 'nibble'.[35]

Bush lawyer (*Rubus australis* and allied species) have fruits consisting of aggregated drupelets similar in form to those of blackberry. They have been eaten raw, stewed, and as jam.[10]

bush lawyer (*Rubus australis*) **Rosaceae**
Engraved from a 1769-70 painting by Sydney Parkinson

PETALS, TEPALS, AND ARILS

Tutu (*Coriaria* species) has enlarged succulent petals around the ripe seeds. Reverend Richard Taylor recorded in his journal that 'At Tawitinui . . . the natives regaled my companions with large quantities of the juice of the tutu boiled with seaweed. This makes a jelly and I think very palatable'. The poisonous seeds were removed by straining the ripe fruit in a basket lined with toetoe (*Cortaderia*) heads.[36]

Pōhuehue (*Muehlenbeckia complexa*) has enlarged white fleshy tepals around the seed, too small to be of much interest. Snowberry (*Gaultheria antipoda* and allied species) has an enlarged fleshy calyx investing the seed capsule, again too small to provide much food.

Tītoki (*Alectryon excelsus*) has a fleshy scarlet aril sitting in the capsule beneath the black seed. It was eaten by the Māori in times of food shortages, although Crowe found the taste to be too astringent to be pleasant.[10]

Most edible native fruits are too small to be used today. However, many prized cultivated fruits were developed from equally dubious progenitors. There is scope here for plant breeders to develop new and interesting New Zealand foods by improving the size and flavour of some of these fruits.

8.4 Beverages

TEA

George Forster in *A Voyage Round the World* 1777, wrote that

> on the first day of our arrival [at Dusky Bay] we found a beautiful tree in flower, something related to the Myrtle genus, of which an infusion had been drank instead of tea on Captain Cook's former voyage. We immediately repeated the experiment with great eagerness as we had not seen any plant which was fit to be used at our table. Its leaves were finely aromatic, astringent, and had a pleasant flavour at the first infusion, but this fine taste went off at the next filling of the teapot, and a great degree of bitterness was then extracted. We never therefore suffered it to be twice infused. The use of this plant, which became general among our crew, probably contributed greatly to restore their strength, and to remove all scorbutic symptoms . . . [18]

The tea plant had been known to Banks and Solander as *Philadelphus*; it is now known as mānuka (*Leptospermum scoparium*). Leaves of bidibid (*Acaena* species), bark of tawa (*Beilschmiedia tawa*), and leaves of kawakawa (*Macropiper excelsum*) have all been used to make tea. Crowe described the taste of bidibid tea as like a mixture of musty tea and cabbage water, and recommended that a little honey be taken with it.[10]

BEER

Captain Cook in his account of the stay at Dusky Bay during the second voyage, wrote that

> the most considerable tree for size is the Spruce-tree, many individuals of which were observed from six to eight to ten feet in girth, and sixty and eighty, to even one hundred feet high, quite large enough to make a main-mast for a fifty-four gun ship.

Of the leaves of this tree Cook's men made beer,

> and when well prepared, and corrected from its extreme astringency by a decoction of Philadelphus or Tea-Plant, proved a good antiscorbutic, and was acknowledged to be little inferior to the American Spruce Beer by those who had experience of both.[37]

The 'Spruce-trees' of Dusky Bay were rimu (*Dacrydium cupressinum*). The 'Tea-Plant' is known today as mānuka (*Leptospermum scoparium*). Taylor (1870) stated that kahikatea (*Dacrycarpus dacrydioides*) was also used instead of rimu.[16]

Gluckman[11] conducted experiments with gourds to find that yeast fermentation produced a most unpalatable drink. He suggested that the old-time Māori had no fermented drink because they lacked a suitable container in which to brew it. He described modern homebrews based on kūmara, potato, and maize; and Crowe[10] listed similar brews made from the baked roots of cabbage tree (*Cordyline australis*), the leaves and/or bark of kohekohe (*Dysoxylum spectabile*), the leaves of kawakawa (*Macropiper excelsum*), and also the roots of supplejack (*Ripogonum scandens*).

Thomas Kirk (1889) described matāī beer:

Matai [*Prumnopitys taxifolia*] is occasionally liable to heart-shakes, but they are rarely serious. When shakes occur the watery portion of the sap accumulates in the cavities in large quantities and during the growing season forms a refreshing beverage for the thirsty bushman, who frequently taps an old tree to procure it. It is rarely obtained in quantity from young trees, and in very old trees has a sour, acrid and unpleasant flavour. When newly tapped on a hot day it is brisk and refreshing; but like many other beverages, speedily becomes 'flat' when exposed to the air.[38]

WINE

Missionaries and early settlers made wine from the fruit of tutu (*Coriaria* species) and makomako or wineberry (*Aristotelia serrata*). A recipe for tutu wine, published in the *Weekly News* (1870) included instructions for bruising the tutu berries as though they were grapes. About six inches of green fern is trampled down in the bottom of a hogshead to act as a 'hop-back' or filter, and the bruised berries are baled in, covered with a sack, and left to ferment. After about ten days, depending on the air temperature, fermentation subsides. (Apply an ear to the side of the hogshead — do not disturb it.) Rack off the liquid into a clean cask. Put the bung in lightly for three months, then secure it. Lift the vent peg in the spring and fall. Bottle after three years. The recipe concluded with a note that the last bottle of a tutu wine made in 1855 was drunk in 1862.[67] Kirk (1889) noted that tutu wine, after standing some time, resembled a light claret. It must be emphasized that tutu is a poisonous plant, and that the seeds and young shoots are particularly dangerous.[38]

COFFEE

J.C. Crawford, a member of the Wellington Philosophical Society, roasted and ground the seeds of karamū (*Coprosma lucida*) and of taupata (*C. repens*) to make coffee for members at a meeting about 1877. *Coprosma* belongs to the

Rubiaceae, the family of the coffee plant, *Coffea arabica*. However, *Coprosma* seeds are much smaller than those of *Coffea* and not worth the trouble of collecting and cleaning.[39] The root of the introduced dandelion (*Taraxacum officinale*) has been dried and ground as a coffee substitute and the native species (*T. magellanicum*) appears suitable for the purpose. The seeds of New Zealand flax (*Phormium tenax*) have also been roasted and ground as a coffee substitute.[40]

BUSH LEMONADE

The juices from rātā vine (*Metrosideros robusta* and allied species), and from bush lawyer (*Rubus australis* and allied species) have been collected and drunk. Young shoots of supplejack (*Ripogonum scandens*) have been eaten and branches of kōtukutuku (*Fuchsia excorticata*) have been cut and sucked to quench thirst. Juice from fleshy bracts of the spadices of kiekie (*Freycinetia banksii*) was mixed with karengo and other edible kinds of seaweed to make a drinkable jelly. Similarly, the dark purple juice from the flowers of tutu, mentioned above, was sometimes mixed with seaweed to make a refreshing drink. The sweet cooked root of cabbage tree (*Cordyline* species) was broken up, cleaned of fibre, and made into a drink or used as a sweetener.

TĪTOKI LIQUEUR

The most important development in the use of native plants in recent years was the launching of a new liqueur based on the fruit of tītoki (*Alectryon excelsus*). The necessary research was carried out by the DSIR team of Horticulture Division at Mount Albert Research Station.

8.5 Seaweeds

During the 1939-45 war, a quantity of karengo (*Porphyra columbina*) was gathered from the New Zealand coast, and sent to the Maori Battalion in the Middle East. V.J. and D.J. Chapman (1980) stated that it was more thirst-quenching on desert marches than chewing gum. Possibly it also had a laxative effect, to keep the army moving.[41]

Lucy Moore (1941, 1944b) and Andrew Crowe (1981) described and discussed seaweeds to be eaten raw, to be cooked as 'greens', and to be added to stews and soups as thickening agents. Sea lettuce (*Ulva lactuca*), recommended as edible raw and cooked, sometimes indicates sewage pollution and should be collected only from clean coastal areas. All seaweeds gathered should be tried with caution, as they are laxative.[10,42,42a]

Y. Elder (1980) described cooking with bull kelp (*Durvillaea antarctica*) to give a flavour to fish, and Lucy Cranwell once described the use of seaweeds at Bethell's guest house to flavour foods in the earth oven or hāngī.[43,44]

In Chapter 13 we mention that seaweeds are an excellent source of food supplements, being rich in minerals and vitamins. Marine algal carbohydrates cannot be digested by human intestinal enzymes and provide the human system with no calories. Edible seaweeds are low in fats, and although their proteins are of better quality than those of higher land plants, they are less digestible. Consequently, it seems that the future role of seaweeds in nutrition will be to contribute to health in such areas as weight reduction. It must be emphasized that overseas workers have examined only a few species of edible algae in the world.[45,46]

8.6 Fungi

The Māori people used about a dozen edible fungi in pre-European times; for example, Best (1942) recorded that the puffball or pukurau (*Calvatia caelata*) was steamed in the earth oven and eaten. Andrew Crowe (1981) gave an account of New Zealand fungi as food, listing sixteen Māori names for different kinds. James Beever prepared *A Dictionary of Maori Plant Names* for the Auckland Botanical Society in 1987 and listed fifty-one Māori names for twelve kinds of fungi. However, our knowledge of what fungi were eaten by the old-time Māori is slight.[10,33,47]

The principal New Zealand fungi of commerce were ear fungus (*Auricularia polytricha*) and the vegetable caterpillar fungus (*Cordyceps* species) mentioned in Chapter 9. We assume that the strains of the common mushroom (*Agaricus campestris*) which are sold commercially are not native to New Zealand (Plate 7).

Ken Cumberland (1981) has told the story of Chew Chong of Canton who came to the North Island about the 1870s by way of the Australian and Otago goldfields. In Taranaki he organized the first butter factory, the sale of butter, and the collection of ear fungus. H.G. Philpott (1937) recorded the history of Chew Chong's pioneer dairy factory. Chew offered threepence a pound for the fungus and sold it to a Chinese merchant in Dunedin. He paid a commission to families who organized the collection of the fungus.[48,49]

About 1889, Thomas Kirk forwarded a sample of *Auricularia polytricha* to Kew, where A.H. Church made an analysis, published in the *Kew Bulletin of Miscellaneous Information* for October 1890, together with some notes which William Colenso had sent to the *Transactions of the Penzance Natural History and Antiquarian Society for 1884-1885*. (The Colensos were an old Penzance family).[50]

Colenso recorded that the only market for ear fungus was China, and that a very minor export trade had grown to 187 tons valued at 8192 pounds sterling in 1881, 339 tons valued at 15,581 pounds in 1882, and 250 tons valued at 9300 pounds in 1883. The trade was due to the enterprise of Chew Chong, and exports have continued in a minor way until recent times. Nothing seems to be known of the trade in vegetable caterpillars, save that they have a role in Chinese medicine. They were used by the pre-European Māori in tattooing.

Cyttaria is a genus of fungi found in South America, Australia, and New Zealand. The three local species are parasitic on silver beech (*Nothofagus menziesii*) and produce galls and coloured fruiting bodies. In South America the galls are used for ornaments and the fruiting bodies for food.[51]

Crowe has several modern analyses of edible fungi, and there are three (two for *Agaricus campestris* and one for 'other edible species, raw') in the *Handbook of the Nutritional Contents of Food*, prepared for the US Department of Agriculture, 1975. Like the edible seaweeds, fungi used for food are poor in carbohydrates, proteins, fats, and calories, although they are comparatively rich in minerals and vitamins. Consequently, the role of edible fungi would seem to be as food supplements, for their minerals and vitamins.[10,52]

Overseas, numerous kinds of fungi are consumed. Over 100 species are used in Britain, and thirty are harvested and sold for food in France. In New Zealand, there is scope for studies of the edibility of native fungi and for publicity to be given to the results. As a beginning, may we draw attention to Andrew Crowe's interesting book and to Marie Taylor's superbly illustrated *Mushrooms and Toadstools in New Zealand*.[10,53]

8.7 Pollen and Honeys

POLLEN

An unusual plant food of the Māori was made from the pollen of raupō or bullrush (*Typha orientalis*). The pollen, gathered from the upper male portion of the inflorescence or collected from the surface of the water, was made into a kind of gruel, or baked into a cake, sometimes flavoured with crushed mānuka beetles (*Pyronota festiva*). D. Burton published modern recipes using raupō pollen.[12]

Overseas, pollen is used to make animal foods, health foods, face creams, toothpastes, and fitness pills. One consumer of the pills is Ronald Reagan, according to an article by C. Newman in the *National Geographic* for October 1984. Fitness pills containing local pollens were analysed by Neville Moar, Botany Division, DSIR, in 1981 and found to contain 10% of exotic plant pollen and 30% of native plant pollen (*Leptospermum, Knightia, Coprosma*, and so on).[54,55]

FLORAL HONEY

In 1947, W.F. Harris and D.W. Filmer, Botany Division, DSIR, made the first identification of pollens in New Zealand honeys and in bee loads. In the early

1980s, Moar and co-workers at Botany Division made a three-year survey of 119 samples of New Zealand honey. The aim was to provide apiarists with guide lines for monitoring the quality of their products, as well as enabling them to meet the requirements for a microscopical examination sometimes demanded by an importing country.[56,57]

Moar (1985) identified 120 pollen types from native and introduced flowering plants, plus a few fern spores in New Zealand honey. The term 'type' is used, as it is often impossible to identify pollen to a particular species. Only a few of the 120 types were consistently predominant; most were secondary or minor components of honey.

White clover (*Trifolium repens*) is the most important nectar source available to bees in New Zealand, and its pollen occurs in honeys as a dominant, secondary, or minor type. The frequency of its occurrence reflects the significant role white clover plays in New Zealand pastures.

Some New Zealand honey is named in the field, on the basis of flowers worked by the bees, and the flavour, odour, and colour of the honey, as 'mānuka', 'pōhutukawa', 'rātā', 'kāmahi', etc. These products sell well locally, are popular with tourists, and are exported successfully to markets overseas, specialist honeys in attractive containers earning a premium price.

'Mānuka' honey has a strong flavour, is usually darker in colour than clover honey, and is produced from mānuka scrub and kānuka bush. Pollen grains of mānuka and kānuka are predominant in the honey, and are too similar to separate.

The term 'rātā' is applied to honey from northern rātā (*Metrosideros robusta*) and southern rātā (*M. umbellata*), and rātā honey is difficult to separate from pōhutukawa honey (*M. excelsa*). Other native plant honeys examined by Moar included 'kāmahi' (*Weinmannia racemosa* and *W. silvicola*), 'rewarewa' (*Knightia excelsa*) and 'matagouri' (*Discaria toumatou*).

Before Neville Moar retired in 1987, he analysed the pollen in several samples of floral honey in which trace organics had been identified by GC-MS analysis in the Chemistry Department of the University of Waikato. Results from the pollen and trace organics analyses were very close, and it is probable that the two techniques will be used together when GC-MS analysis is developed.[68]

In August 1990, the University of Waikato invited applications for a two-year postdoctoral fellowship to develop GC-MS analysis of trace organics in New Zealand floral honeys, and the successful applicant is expected to begin work early in 1991. Meantime, progress has been slowed by the lack of a replacement for Neville Moar to do pollen analyses of honey samples and by delay in the publication of his atlas of New Zealand pollens. However, the Lottery Board granted funds for its publication in November 1990, and it is hoped that the atlas will be available in 1991.[66]

Honey is a very old remedy for bacterial infections and, as more bacteria seem to be developing resistance to antibiotics, some medical practitioners are turning to it. P.C. Molan, I.M. Smith and G.M. Reid (1988) found that the most active antibacterial honeys available in New Zealand are kānuka (*Kunzea ericoides*) mānuka (*Leptospermum scoparium*), and pennyroyal (*Mentha pulegium*). Nodding thistle (*Carduus nutans*), kāmahi (*Weinmannia racemosa*), and the buttercup (*Ranunculus* species) have lower activity. Rewarewa (*Knightia excelsa*), clover (*Trifolium* species), heather (*Erica lusitanica*) tāwari (*Ixerba brexioides*), rātā (*Metrosideros* species), tōwai (*Weinmannia silvicola*), thyme (*Thymus vulgaris*), and blue borage (*Echium vulgaris*) honeys have the lowest antibacterial activity. Pennyroyal, nodding thistle, clover, heather, thyme, and blue borage are introduced species; kānuka, mānuka, kāmahi, some buttercups, rewarewa, tāwari, rātā and tōwai are native plants.[64]

Antibacterial activity in mānuka honey is due to a group of aromatic acids related to the commonly used food preservative sodium benzoate. Presumably they occur in particularly high concentrations in mānuka nectar.[65] In further studies, Peter Molan and his colleagues at the University of Waikato have demonstrated that the antibacterial activity of mānuka honey is due to substances other than hydrogen peroxide and is relatively stable. However, anyone wanting to use mānuka honey as an antibacterial substance should obtain it as fresh as possible, unheated, and stored out of light.[62-65]

HONEYDEW HONEY

Honeydew honey is produced by bees feeding on honeydew which is excreted by a soft scale insect (*Ultracoelostoma assimile*). The scale insect lives in waxy capsules beneath the bark of beech trees (*Nothofagus species*), except silver beech (*N. menziesii*). It feeds on the phloem sap and drains its wastes and unused sap to the exterior of the tree by a hollow waxy thread. A droplet of this sugary substance, known as honeydew, hangs on the tip of each anal tube of second and third instars. Honeydew nourishes black sooty mould fungi, insects, nectar feeding birds, and is the source of honeydew honey.[58]

The beech scale insects are widespread in the South Island — on the West Coast north of Greymouth and on the east coast north of Mt Somers, in Marlborough and in Nelson. Infestation is particularly heavy in the Canterbury foothill forests.

Honeydew honey is amber coloured and strongly flavoured. It can be distinguished from floral honey by its higher content of ash, dextrin, sucrose, reducing sugars, higher pH, and greater free acidity. Spores and hyphae of the sooty mould and beech pollen grains are also present.[59]

Honeydew honey production in 1989 was estimated as 342 tonnes in Canterbury (MAF apiary report, unpublished) and a small quantity in the

Murchison area (apiarist pers. comm.). A large proportion is exported to Germany where it is generally blended with their darker honeys. Wasps are the greatest harvesters of honeydew in summer and autumn in Nelson forests. In a recent Nelson study (Moller and Tilley, 1989) wasps are thought to have a strong dampening effect on the amount of honeydew available between February and May. They also are known to feed on native and introduced insects, including bees, reducing honeydew honey production further and possibly affecting insectivorous birds.[60]

1. Cheeseman, T.F. (1901) 33 *Trans. N.Z. Inst.* 306.
2. Fankhauser, B.L. 1986. Ph.D. Thesis, University of Otago.
3. Fankhauser, B.L. 'The Maori use of ti (cabbage trees) for food' In Harris, W., Kapoor, P. (eds) *Nga Mahi Maori O Te Wao Nui A Tane. Contributions to an International Workshop on Ethnobotany* Christchurch (DSIR Botany Division) 1990 209p. p. 43.
4. Fankhauser, B.L., Brasch, D.J. (1985) 1 *N.Z.J. Tech.* 27.
5. Munro, M.H.G. Pers. commun. to Brooker, S.G.
6. Mann, J., Stewart, K. 'Ti (cabbage tree) fructose'. In Harris, W. and Kapoor, P. (eds) *Nga Mahi Maori O Te Wao Nui A Tane: Contributions to an International Workshop on Ethnobotany.* Christchurch (DSIR Botany Div.) 1990 209p. p. 182.
7. Thomson, A.D. (1989) 124 *Botany Division Newsletter* 10.
8. Thomson, A.D. (1990) 130 *Botany Division Newsletter* 8.
9. Smith, A.M., Fenton, E.W. (1944) 63 *J. Soc. Chem. Ind. London* 218.
10. Crowe, A. *A Field Guide to the Native Edible Plants of New Zealand* Auckland (Collins) 1981 194p.
11. Gluckman, L.K. *Medical History of New Zealand prior to 1860* (Tangiwai on cover). Auckland (published by the author) 1976 271p.
12. Burton, D. *Two Hundred Years of New Zealand Food and Cookery* Wellington (Reed) 1982 168p.
13. Houghton, P. *The First New Zealanders* Auckland (Hodder & Stoughton) 1980 156p.
14. Nicholl, R. Pers. commun. to Brooker, S.G.
15. Colenso, W. (1881) 13 *Trans. N. Z. Inst.* 3.
16. Taylor, R. *Te Ika A Maui or New Zealand and its Inhabitants* London (Wetheim & Macintosh) and Wanganui (Jones) 1855, 490p; 2nd ed. 1870, 713p. Reprinted Wellington (Reed) 1974.
17. Hooker, J.D. (ed.) *Journal of the Rt. Hon. Sir Joseph Banks, Bart. K.B., P.R.S., during Captain Cook's First voyage in H.M.S. Endeavour in 1768-1771 etc..* London (Macmillan) 1896 466p.
18. Forster, G. *A Voyage round the World in His Majesty's Sloop Resolution* London (White) 1777 2 vols.
19. Colenso, W. (1869, 2nd ed. 1875) 1 Trans N.Z. Inst. 233.
20. Ventenat, E.P. *Jardin de la Malmaison.* Paris (Herhan) 1804. Vol. 2, tab. 81. *Apium prostratum.*
21. Forster, G. *De Plantis Esculentis Insularum Oceani Australis Commentatio Botanica* Berlin (Haude & Spener) 1786 80p.
22. Brooker, S.G., Cambie, R.C., Cooper, R.C. *New Zealand Medicinal Plants* Auckland (Heinemann) 1987 268p.

114   *New Zealand's Economic Native Plants*

23. Webb, C.J., Sykes, W.R., Garnock-Jones, P.J. *Flora of New Zealand Volume IV. Naturalised Pteridophytes, Gymnosperms, Dicotyledons* Christchurch (DSIR Botany Division) 1988 1365p.
24. Wilson, C.M., Given, D.R. *Threatened Plants of New Zealand* Wellington (DSIR Publishing) 1989 151p.
25. Moore, L.B., Irwin, J.B. *The Oxford Book of New Zealand Plants* Wellington (Oxford University Press) 1978 234p.
26. Anderson, J. (1821) 4 *Trans. Hort. Soc. London* 488.
27. Morrell, W.P. (ed.) *Sir Joseph Banks in New Zealand; from his Journal 1769-1770* Wellington (Reed) 1958 159p.
28. Bellinghausen, T. 'Russians at Macquarie Island, 1820' in McNab, R. *Murihiku and the Southern Islands* Invercargill (W. Smith) 1907 377p, p. 190. Facsimile edition by Wilson & Horton.
29. Mueller, F. von *Select Extra-tropical Plants* Sydney (Government Printer) 1881 403p. (and many other editions).
30. Garnock-Jones, P.J., Johnson, P. (1987) 25 *N.Z. J. Bot.* 603.
31. Crayon, (1842) 15 *New Zealand Journal, London* 278.
32. Drummond, A. (ed.) *The Auckland Journals of Vicesimus Lush* 1850-63 Christchurch (Pegasus Press) 1971 280p.
33. Best, E. *Forest Lore of the Maori* Wellington (Government Printer) 1942 420p.
34. Molloy, B. 'The origin, relationships, and use of karaka or kopi (*Corynocarpus laevigatus*) in Harris, W. and Kapoor, P. (eds) *Nga Mahi Maori O Te Wao Nui A Tane. Contributions to an International Workshop on Ethnobotany* Christchurch (DSIR Botany Div.) 1990 209p, p.48.
35. Hooker, J.D. *Handbook of the New Zealand Flora* London (Lovell Reeve) 1864 vol. 1, 392p; 1867 Vol. 2, p. 393-798.
36. Taylor, R. 1833-1873 Journals. Typescript copies in the Library, Auckland Institute and Museum.
37. Cook, J. *A Journey towards the South Pole and round the World. Performed in His Majesty's Ships the Resolution and Adventure, in the years 1772, 1773, 1774, 1775* London (Strahan & Cadell) 1777 2 vols xi 378 396p.
38. Kirk, T. *The Forest Flora of New Zealand* Wellington (Government Printer) 1889 345p. 142pl.
39. Crawford, J.C. (1877) 9 *Trans. N.Z. Inst.* 545.
40. Kirk, T., *c* 1870 'Plants indigenous to New Zealand available for food, medicine, and other economic purposes, revised by T. Kirk'. Typed manuscript in Library, Auckland Institute and Museum.
41. Chapman, V.J., Chapman, D.J. *Seaweeds and their Uses* London (Chapman & Hall) 1980 3rd ed. 334p.
42. Moore, L.B. *The Economic Importance of Seaweeds* Wellington (Government Printer) 1941; DSIR Bull. 85, 40p. Reprinted 1966; Moore, L.B. (1944) 7(3) Bull. N.Z. Women's Food Value League 12.
43. Elder, Y. (1980) 14 *Canterbury Bot. Soc. J.* 54.
44. Cranwell, L. *The Botany of Auckland. A Book for All Seasons* Auckland Institute and War Memorial Museum Handbook 1981 156p.
45. Arasaki, S., Arasaki, T. *Low Calorie, High Nutrition Vegetables From the Sea* Tokyo (Japan Pubs. Inc.) 1983 196p.
46. Fujiwara-Arasaki, T., Mino, N., Kuroda, M. (1984) 116/117 *Hydrobiologia* 513.
47. Beever, J. *Dictionary of Maori Plant Names* Auckland 1987 Bot. Soc. Bull. 16, 74p.

48. Cumberland, K.B. *Landmarks* Surry Hills, N.S.W. (Readers Digest) 1981 304p.
49. Philpott, H.G. *A History of the New Zealand Dairy Industry 1840-1935* Wellington (Government Printer) 1937 413p.
50. Church, A.H. (1890) *Kew Bull. Misc. Information* 46, including notes by Colenso, W.
51. Wardle, J.A. 'Silver Beech' In Knox, R. (Ed.) *New Zealand's Nature Heritage* Wellington (Paul Hamlyn) 1974. 6 vols issued in weekly parts. 2948p. with endpapers for each part. p. 1011.
52. US Department of Agriculture *Handbook of the Nutritional Contents of Foods* New York (Dover) 1975 190p.
53. Taylor, M. *Mushrooms and Toadstools in New Zealand* Wellington (Reed) 1970 32p.
54. Newman, C. (1984) 166(4) *National Geographic* 490.
55. Moar, N. (1981) 72 *Botany Division Newsletter* 7.
56. Harris, W.F., Filmer, D.W. (1948) 30A *N.Z.J. Sci. Tech.* 178.
57. Moar, N.T. (1985) 28 *N.Z.J. Agric. Res.* 39.
58. Crozier, L.R. (1981) 26 *N.Z.J. Forestry* 200.
59. Kirkwood, K.C., Mitchell, T.J., Smith, D. (1960) 85 *The Analyst* 412.
60. Moller, H., Tilley, J.A.V. (1989) 16 *N.Z.J. Zoology* 289.
61. Rees-George, J., Robertson, G., Hawthorne, B. (1990) 28 *N.Z.J. Bot.* 363.
62. Molan, P.C. (1985 Summer) *The New Zealand Beekeeper* 29.
63. Molan, P.C., Russell, K.M., (1988) 27(1) *J. Apicultural Res.* 62.
64. Molan, P.C., Smith, I.M., Reid, G.M. (1988) 27(4) *J. Apicultural Res.* 252.
65. Russell, K.M., Molan, P.C., Wilkins, A.L., Holland, P.T. (1990) 38 *J. Agric. Food Chem.* 10.
66. Dr Neville Moar, 8 Maurice Street, Lincoln. Letter of 8 November 1990.
67. Anon. (1870) *Weekly News* 15 January, p. 12, col. 2.
68. Dr A. Wilkins, Chemistry Department, University of Waikato, pers. commun. November 1990.

9

Potions and poisons

9.1 Medicinal Plants

One of the most important achievements of early nineteenth century chemists was the isolation of the active principles from a number of plant sources which had long been known for their medicinal properties. Examples include morphine from the dried latex of opium poppies (1805), strychnine from *Nux vomica* seeds (1819), quinine from cinchona bark (1820), and reserpine from the Indian plant *Rauwolfia serpentina* (1952). For scientists interested in the chemistry and pharmacology of plant products, a knowledge of the uses of local plants by the indigenous population can serve as a guide for the further investigation of plant species. In many countries, notably in India, Malaysia, and South Africa, these facts have been collected and published in authoritative works of reference. *New Zealand Medicinal Plants*[1] gives a comprehensive record of the medicinal uses of New Zealand plants dating from the time of Captain Cook's exploration of 1769. Gluckman[2] makes several references to the use of medicinal plants in a valuable book on the medical history of New Zealand up until 1860.

The purpose of this chapter is to indicate to what extent Māori and early European settlers in New Zealand used plants as medication, and to deal more closely with those plants which appear to have been of particular value. We also mention some of the chemical constituents of New Zealand plants, and indicate how some of them might possibly be utilized for medicinal purposes. For a full, detailed discussion on the use of native plants in medicine in New Zealand, together with some related pharmacology, the reader is referred to our earlier book.[1]

There is some doubt about the internal use of plants in medicine by the early Māori (see below) and thus the earliest use of New Zealand plants for medicinal purposes was perhaps by Captain Cook in 1769. Cook realized the value of fresh vegetables in preventing scurvy, a common ailment of sailors on long voyages in the days of sailing ships. Wherever he went, he tried out plants as antiscorbutics, and used to good advantage New Zealand spinach (*Tetragonia tetragonioides*) which has subsequently been shown to contain bctain, alkaloids, saponins,[3] and tetragonin, a yeast growth regulator.[4] The plant also exhibits carbonic anhydrase activity,[5] and thus apparently has the opposite effects of the sulphonamide drugs. More recently the plant has been shown to contain cerebrosides, compounds which have anti-ulcerogenic properties.[6] Cook also collected boat-loads of 'scurvy grass' (*Lepidium oleraceum*). This latter plant is related to watercress which is a well-known antiscorbutic, presumably because of the presence of vitamin C, or ascorbic acid (see fig. formula 1). Cook records how he made a 'Sour Kroutt' from these plants; in order to get his men to eat the concoction, he initially allowed only his officers to have it. Within a week, there was such a demand for it from the lower deck that it had to be rationed.[7] 'Scurvy grass' has now almost disappeared from the New Zealand flora, largely destroyed by introduced animals.

The French explorer, de Surville, who visited New Zealand about the same time as Cook, had similar interests. His lieutenant, Pottier L'Horme, recorded the use of New Zealand plants for the same purpose.[8] They were not clearly identified, but one was possibly Cook's scurvy grass; another, because of reference to the acid milk in the stem, was probably sow thistle — the 'pūwhā' of the Māori. We have referred to the commercial cultivation of 'pūwhā' in Chapter 8.

It is not clear what use the Māori made of plants as medicine in pre-European times. Elsdon Best considered that the 'Maori of old relied principally upon his priest when attacked by sickness, and that the priests did not deal in simples or herbal remedies . . .'[9-11] The tohunga used various plants in dealing with sick people, but these uses were more in the nature of religious rites to exorcise demons than medicinal. However, Cook recorded the use of 'cellery' placed on hot stones to provide healing vapour baths. Other early Europeans in New Zealand, for example J. Rutherford, [12] who lived with a Māori tribe from 1816 to 1826, recorded the use of plants in Māori medicine, as did Major R.A. Cruise, [13] who spent ten months in thc country in 1820, and J. Crocombe, [14] who practised as a doctor at a Māori and whaling settlement in Otago from 1838. William Colenso, [15] a mission printer who arrived at Paihia from Sydney in 1834 and spent much time with the Māori, came to the conclusion that their practice was to use medicinal vapour baths and herbal remedies for external complaints only.

Any precise conclusion on this issue is hampered by the fact that Māori copied European from the earliest contact. European explorers and settlers, often far from professional medical aid, were more or less forced to treat common ailments such as dysentery, constipation, and colds, as well as injuries, with whatever native plants came to hand. The Māori enthusiastically followed the practice of the newcomers in experimenting with the medicinal uses of various native plants. As a result, many Māori remedies or 'bush cures' have arisen.

The missionaries William Williams (1826),[16] Richard Taylor (1844-1852),[17] Charles Baker (1854),[18] Thomas Chapman (1857),[19] and William Puckey (1858),[20] all recorded the enthusiastic adoption of internal medicines by various tribes, several of whom experimented bravely with their own concoctions.

More important in the medical sense is the work of James F. Neil, who published his *New Zealand Herb Doctor* in 1889.[21] It is a very interesting work, where the author introduces in several places his Christian faith, and safeguards himself from any failure of his nostrums by saying 'It is but right for us as to the means, but we should never forget that life and death are in the hands of our Creator' in a paragraph which closes with a devout 'Amen'. Neil, who had attended 'three sessions' at Otago Medical School, and colleges in Chicago and New York, included eighteen native plants in his herbal: koromiko, New Zealand flax, mānuka, hīnau, karaka, supplejack (*Ripogonum*), kawakawa, kohekohe, kōhia, kopakopa (plantain), rātā, pukatea, horopito, pōhutukawa, tānekaha, tutu, tikumu (*Celmisia*), and spear-grass (*Aciphylla*).

A later and more authoritative compilation is that of Goldie published in 1905.[22] In this work, Goldie had the active co-operation of Elsdon Best, and listed over 100 plants of reputed value.

One of the most important of the 'bush cures' is the use of *Hebe* species, particularly koromiko (*Hebe salicifolia*, *H. stricta*, and allied species), which is a well-known and authenticated remedy for dysentery.[1] Koromiko is the only native plant to have received recognition in British medicine, being listed as a remedy for diarrhoea in the 1895 Extra Pharmacopoeia, London. It is still valued as a remedy for dysentery and diarrhoea and, during World War II, quantities of the dried plant were sent to the North African front, where it was used effectively by Māori troops. The value of the plant has been traced to its anti-peristaltic action. The active principle was originally suggested to be the tannins which are present in the plant, but more recent work by Martin-Smith has indicated that it is a phenolic glycoside.[23] Even so, it is somewhat surprising that so little work has been carried out on this plant.

One or two native fungi have been, or may be, used in herbal medicine overseas. Chinese merchants in New Zealand have bought ear fungus (*Auricularia polytricha*) for use in medicine, and an allied species is said 'to strengthen the lungs and activate blood circulation', and to be used for a variety of ailments ranging from traumatic injuries to bleeding haemorrhoids. The vegetable caterpillar, a fungus (*Cordyceps* species) which is parasitic on some insect pupae, is also used in Chinese medicine and is said to be very highly prized. From time to time, it is found in numbers in the New Zealand bush, and its potential for export merits further inquiry.

Impinging on the area covered by this chapter is the growing interest in 'health' foods, which are sold in special shops for their 'medicinal virtues'. In view of the large number of New Zealand plants reported to have some value as herbal remedies, it is somewhat disappointing to report that few of the large number of products offered for sale in these emporia are of local origin, even when the plant is readily obtainable. As an example, one so-called 'health' shop offers kelp of Australian origin. This plant has been found by the New Zealand DSIR to contain useful vitamins and, like most seaweeds, is a good source of iodine. There is no reason why it should be imported in view of the ready availability of New Zealand kelp.

Some years ago, there were a number of herbalists offering free advice to people with ailments, and making a living by the sale of various natural products with putative therapeutic powers. One of the most prominent was Rev. Edgar Ward, who qualified as a pharmacist before entering the Anglican ministry. Later he became vicar of Kaitaia, where he was concerned about the number of Māori children who had contracted tuberculosis. He administered

Reverend Edgar Ward
From neg. *c.* 1738, Library, Auckland Institute and Museum

a mixture of kūmarahou (*Pomaderris kumeraho*) and other herbs with some success. He called this nostrum 'Kuranui', a name not in the Māori dictionaries, but having the right sound. He subsequently left the ministry and set up as a herbalist in Auckland, where he practised for many years.[24]

Mother Mary Aubert was another herbalist who gained wide publicity and respect, particularly within her own Catholic church, whose official newsletter contained testimonials from the Catholic hierarchy. Her medicines were given names such as Natanata (for stomach troubles), Marupa (for throat and bronchial ailments), Parano, and Karana, redolent of Māori and of her native French. She was a most remarkable woman, whose social and missionary work reflected much credit on her church, which she served in New Zealand for over fifty years. For a large part of this time she operated from the mission station at Jerusalem on the Wanganui River. The proceeds of the sale of her potions were used to support novices in the early years of their profession. The preparation and use of these herbal medicines is discussed in some detail by Rafter.[25] In order to support her religious and humanitarian work, a commercial arrangement with Kempthorne, Prosser Co. was made in 1891, but dilution of the remedies by the company resulted in legal proceedings in 1894. Subsequently Mother Aubert's preparations, made up by herself and the Sisters, were sold through Sharland and Co. and sales extended to Australia. However, when she went to Rome in 1913, she destroyed all her recipes — mainly, according to Rafter, because members of the B.M.A. would have refused to help her in her children's home unless she desisted. Samples are preserved, however, at the Home of Compassion, Island Bay, Wellington.

It seems likely that the operations of these herbalists may have been greatly affected by the medical benefits scheme introduced by the Labour Government in 1939, under which both consultations and medicines became free. The 1984 New Zealand budget included a charge on prescriptions, while inflation greatly increased doctor's fees of which social security now pays only a minor proportion. This may in part account for the recent proliferation of 'health' and 'health food' shops.

When New Zealand's drug supplies were threatened during the Second World War, Ruth Mason (Botany Division, DSIR, Wellington) obtained reports on native medicinal plants from Māori schoolteachers, pharmacists, and others. In 1945, Olga Adams wrote a popular account of Māori medicinal plants.[26] Adams was of Māori descent and noted details of preparations not recorded previously. Her work is out of print, but it was included in *Medicines of the Maori* by Christina Macdonald (1973).[27] In 1954, Dr G.B. Palmer[28] described some of the beliefs of the contemporary Māori, and in 1976 Dr L.K. Gluckman[2] reviewed the interaction of Māori and Pākehā disease and medicine prior to 1860.

Of the natural products from any indigenous flora, it is the alkaloids which provide the bulk of constituents that exhibit authenticated medicinal properties. They are usually bitter, may serve as tonics, emetics, or purgatives, and are administered in low concentrations because most are poisonous. Unfortunately, in keeping with the floras of other temperate regions, New

Zealand is not rich in those families, such as the Apocynaceae, which characteristically yield alkaloids. This was borne out by a survey of the native flora for alkaloids, which was carried out some twenty years ago.[29,30] Nevertheless, despite the lack of alkaloid-bearing species, some New Zealand plants containing these compounds have proved to be of definite value.

As a consequence of their commercial potential, major interest has centred on the New Zealand *Solanum* species, poroporo (*Solanum aviculare*), and *S. laciniatum*. Each contains the steroidal alkaloid solasonine, which can be hydrolysed to the aglycone solasodine (2). The latter has a structure similar to that of diosgenin (3), which is obtained from the Mexican yam (*Dioscorea mexicana*), and which is one of the major source materials for the commercial production of steroids used in the pharmaceutical industry.[31] Details of these processes and the use of *Solanum* alkaloids as medicinals in their own right are given in Chapter 12.

One of the few New Zealand plants to have been given more than passing pharmacological examination is pukatea (*Laurelia novae-zelandiae*). The Māori used to chew the leaves of this tree to relieve toothache and for various other ailments, such as syphilis. In recent years, the leaves have been shown to have both local and general analgesic activity and the bark has been shown to contain a series of alkaloids, the major one of which is pukateine (4).[32] Pukateine bears some resemblance to morphine (5) in structure and has been shown to produce substantial relief of pain without the side effects of morphine.[32]

Sometimes the use of a native plant was blended with a religious rite; for example, the juice from kōwhai (*Sophora*) bark was taken as a purgative, and to treat scabies and ringworm, but the bark had to be from a tree growing on a hillside and from a root pointing towards the sun.[1] New Zealand has two main species of *Sophora, viz. S. microphylla* and *S. tetraptera.* Both contain the alkaloids matrine (6) and cytisine (7) in all parts of the plant[32] and it is probable that the purgative properties of the plants are associated with the presence of these compounds.

Mairehau (*Phebalium nudum*), wineberry (*Aristotelia serrata*), and rangiora (*Brachyglottis repanda*), all contain interesting alkaloids, e.g. (8), (9), and (10) respectively,[32] but, although they were used for various medicinal purposes by the Māori, there is no obvious correlation between their medicinal use and the constituent alkaloids.

Tutu (*Coriaria arborea*) does not contain alkaloids, but it was used by the Māori for medicinal purposes, including the treatment of constipation caused by eating other poisonous plants. The active principle is tutin (11), which resembles the convulsant poison, picrotoxinin (12), in both structural and pharmacological properties. Picrotoxinin is used in the treatment of poisoning by depressants of the central nervous system.[1]

Despite the fact that it had no recorded medicinal uses by the Māori, there are many reports of toxicity in livestock, especially horses, of the sprawling shrub *Pimelea prostrata* which was known to the early settlers as Strathmore weed. In recent years, the plant has been shown to contain the tumour-inhibiting compound prostratin (13), which is related to the well-known anti-tumour drug phorbol.[32]

During the mid and late 1950s, extracts from many New Zealand plants were examined for anti-tumour activity by the Auckland Division of the New Zealand Cancer Society under the direction of the late Dr B.F. Cain. About 10% showed some effect, which is about the same as the flora of most countries which have been investigated. The most active compound to arise from this extensive survey was the terphenyl, polyporic acid (14), from the lichen, formerly *Sticta* and now *Pseudocyphellaria coronata* which was shown to have anti-leukaemic activity in mice.[34] Polyporic acid is rather insoluble, and in an attempt to improve the solubility, Cain prepared and tested over 100 of its derivatives. Disappointingly, not one of these compounds approached the natural product in activity. Also, for various reasons, none of the other compounds or extracts from the survey of the native flora could be used and the unit turned its attention towards synthetic compounds.

One plant of interest is ongaonga (*Urtica ferox*), the giant stinging-nettle of New Zealand. This contains a toxin which has killed livestock and at least one human; the extract of just five hairs can kill a guinea pig. Acetylcholine (15), histamine (16), and 5-hydroxytryptamine (17) have been shown to be present in the hairs, but in no greater quantity than in the hairs of the less dangerous common European nettle (*Urtica urens*). Acetylcholine exerts a powerful stimulation of the parasympathetic nerve system; it is also a cardiac depressant and an effective vasodilator. The toxin of *U. ferox* is water soluble, but all attempts to purify the compound have resulted in partial destruction of the toxin. However, it has been separated into two distinct poisonous fractions, one of which has local-anaesthetic activity. The chief toxin has been named 'triffydin', no doubt after the mobile plants of John Wyndham's well-known science fiction novel *The Day of the Triffids*.[1] The plant also contains silicon in an organic form.

Most New Zealand gymnosperms have, at various times, been used for a variety of medicinal purposes ranging from the treatment of baldness to the relief of urinary and other internal complaints. Kahikatea (*Dacrycarpus dacrydioides*) and rimu (*Dacrydium cupressinum*) each contain podocarpic acid(18), which is effective in promoting the flow of bile and whose reduction product, podocarpatrienol, is estrogenic.[35] Considerable pharmacological work has been carried out on podocarpic acid derivatives by Dr R.H. Bible at the Searle Research Laboratories in the US.[36,37] The wood of the related matai (*Prumnopitys taxifolia*) which was tapped by bushmen to obtain juice 'drunk to check the advance of consumption' and used as an antiseptic, contains genistein(19), an estrogenic isoflavone, and large amounts of the lignan matairesinol (20), which has antimitotic properties and anti-tumour activity in mice.[1] Workers at the University of Auckland have converted the dimethyl ether of matairesinol into an isostegane derivative (21), with the same skeleton as the therapeutically important lignan steganacin.[38]

Without a doubt, the New Zealand plant which is mentioned most frequently in earlier reports of medicinal use is native flax (*Phormium tenax*). There are many reports of the use of it as an antiseptic, purgative, and for the treatment of burns. In an emergency, several leaves bound together make a substitute splint when tied on with a twine made from strips of the leaf. Gluckman[2] also quoted a report that newborn Māori babies were washed in water containing gum such as that obtained from flax. There are so many reports about the styptic healing properties and other benefits of the gum found at the base of the leaves that it is difficult to believe it has no virtue. The gum has been investigated by McIlroy, who found it to be mainly composed of an aldobiuronic acid made up of D-xylose and D-glucuronic acid units, but this gives no clue to the reason for its healing effects.[32] There is an intriguing report of the isolation of a red crystalline compound from the rhizome which could possibly be a purgative anthraquinone.[1] Recently, scientists at Chemistry Division, DSIR, have shown that the roots of *Phormium tenax* and *P. cookianum* do in fact contain the anthraquinones chrysophanol and emodin as well as the red compounds dianellidin and stypandrone.[39] Also, the late Professor S.M. Kupchan, of the University of Virginia, followed up an earlier report of the anti-tumour properties of flax, and isolated the active inhibitors, which turned out to be a series of cucurbitacins (22), which are well-known for their anti-tumour properties.[40] A pharmacy in Opunake is currently attempting to produce a medicinal soap incorporating extracts of flax roots, for soothing and whitening reddened, chapped, and inflamed skin.[41]

Another plant which has been popularly believed to have remarkable curative powers is kūmarahou (*Pomaderris kumeraho*), also known as 'gumdigger's soap' because of the frothing properties of the leaves when shaken with water (see Chapter 10). At various times, extracts of the leaves have been claimed as a panacea for most ills, ranging from skin cancer to coughs and colds. For many years, kūmarahou featured in various advertisements for patent medicines — one such infusion figured in a popular herbal remedy for chest complaints. Since this medicine also contained ipecacuanha, a standard cough remedy, and strychnine, which has a reputation as a tonic, the merits of kūmarahou itself are doubtful. It contains the flavonols quercetin (23, R=H) and kaempferol (23, R=OH), as well as the dicoumarin ellagic acid (24) and its O-methyl ethers,[32] but none of these compounds are renowned for their curative properties.

Some New Zealand plants which yield essential oils contain compounds of potential medicinal use. For example, mānuka (*Leptospermum scoparium*) and kānuka (*Kunzea ericoides*) both contain leptospermone (25), a triketone which has anthelmintic (anti-worm) and insecticidal properties. Mānuka could be of commercial value, since the essential oils of other species of the family Myrtaceae are known to be effective against *Staphylococcus aureus*, a common infectious bacterium. Honey made from mānuka contains methyl syringate, and 3,4,5-trimethoxybenzoic acid which have been shown to have antibacterial properties.[42] The oil of kawakawa (*Macropiper excelsum*) contains myristicin (26), which is similar to eugenol (27), the major constituent of cloves — hence the use of the leaves by the Māori as a remedy for toothache and pain had a sound basis (Plate 10). The fern *Asplenium lamprophyllum*, but not other species of this genus, contains methyl salicylate (28) (oil of wintergreen), which produces marked analgesic effects and has antiseptic properties.[1]

spleenwort (*Asplenium lamprophyllum*) **Aspleniaceae**
Photo: J.E. Braggins

One of the most notable investigations carried out on a New Zealand plant with the aim of isolating the medicinal component is a recent investigation of New Zealand seven-finger (*Schefflera digitata*), the sap of which, according to R. Taylor,[43] had been used for curing ringworm and scrofulous sores. Dr J.R.L. Walker at Canterbury University has shown that the active principle is an unsaturated alcohol falcarindiol (29), which shows remarkable specific activity against common dermatophyte fungi such as that which causes ringworm. Falcarindiol is unusual in that, unlike other antifungal agents which attack only the growing mycelium, it acts by inhibiting spore germination.[44,45] This is one of the few cases of a New Zealand native plant where a specific remedial effect has been related to a specific chemical constituent.

Walker and co-workers[46] have also shown that the active principle of horopito (*Pseudowintera axillaris*) is the bicyclic sesquiterpenoid dialdehyde, polygodial (30). The bark of this plant has been used as a substitute for quinine, while a decoction of the leaves, known as 'Māori painkiller', was taken for a variety of complaints including stomach-ache, skin diseases, and venereal diseases. Polygodial exhibits antibiotic activity especially against the yeast *Candida albicans* which is responsible for skin infections and for thrush in infants. *Candida albicans* is a normal constituent of the body flora of humans, but it is not usually pathogenic unless the person is debilitated by disease. Polygodial works by making a hole in cell membranes thus permitting antibiotics to enter and kill bacteria that are otherwise resistant or need massive doses of antibiotic. The compound also makes antibacterials effective against fungal yeast in the same way. This latter property was discovered by a Japanese chemist, I. Kubo, who read a newspaper report of a boy who had been dosed with antibiotics to such an extent that his bacterial balance was disrupted and he was infested with yeast. Every time he ate sugar-containing food he became drunk. His affliction was cured by Kubo who treated the boy with polygodial to make the antibiotics effective against the yeast.[47]

Barks of New Zealand *Coprosma* species have considerable tinctorial properties and are rich in hydroxylated anthraquinones. Some, such as morindone (31) from *C. australis* are substituted in the 2-position like those found as mould metabolites, and are obtained in a yield as high as 20% of the dry weight of the bark.[32] These anthraquinones have purgative properties, and could also have potential for the synthesis of important drugs. Workers at the University of Auckland are currently engaged on the conversion of such anthraquinones into analogues of daunomycinone (32), the aglycone of the anthracycline anti-tumour agent, daunomycin.[48] New Zealand *Coprosma* species also contain asperuloside (33), an iridoid which can be readily isolated, for example, in high yield from *C. australis*.[32] Although asperuloside itself has no noted medicinal properties, attempts have been made at the University of Auckland to convert it into a suitable material for the synthesis of prostaglandins.[49]

Unfortunately, very little pharmacological work has been carried out on New Zealand plants. What little that has been done is mainly due to Professor F.N. Fastier and his colleagues at the Department of Pharmacy, University of Otago.[50] A survey of New Zealand plants for antibiotic activity has recently been published by Dr J.R.L. Walker and co-workers[51] at the University of Canterbury while a research project initiated by Drs K. Markham and S. Bloor, Chemistry Division, DSIR in 1988 aims to survey the native plants of New Zealand for pharmacologically active compounds.[52] The aim of this survey is to identify species showing antifungal, antibacterial, antiviral and anticancer activity from *in vitro* tests. It is intended to pursue those species showing activity to the stage where pure active components can be isolated and identified and if necessary produced in larger quantities for more extensive evaluation and possible commercial exploitation. Several taxa are reported to have shown strong cytotoxic activity.

In recent years there has been a reawakening of interest in the chemical constituents of native plants and in their potential commercial value. New Zealand has a unique and most interesting flora, but although the country has been occupied by Europeans for less than 200 years, its list of endangered species is becoming alarmingly large. It is to be hoped that these and other plants will be examined before they are lost to us forever.

9.2 Poisonous Plants

Many poisonous plants have an important effect on the livestock industry in New Zealand, and thus are mentioned here in a negative sense, since losses by sickness and death due to plant poisoning can total a considerable sum. Poisonous plants in New Zealand have been comprehensively reviewed by Dr H.E. Connor, Botany Division, DSIR, Lincoln[53] who has shown that by far the majority of poisonous plants are not indigenous to this country.

The most important poisonous native plants include bracken, facial eczema fungus, karaka, ngaio, ongaonga, pinātoro, tutu and waoriki. Algal blooms on farm ponds, which can cause stock losses, are discussed in the chapter on freshwater algae; red tides that have resulted in fish and shellfish deaths are described in the chapter on marine algae.

Bracken (*Pteridium esculentum*) is poisonous to cattle, horses, and sheep, and is reputed to be a cause of stomach cancer in humans. Evans and Osman[54] isolated shikimic acid from bracken and attributed carcinogenic and mutagenic properties to it; they also indicated that bracken contains another carcinogen stronger than shikimic acid. Bracken also has an enzyme thiaminase which destroys thiamine; this results in a Vitamin B deficiency which causes serious poisoning, particularly in horses and cattle (see Connor[53] for a discussion of bracken poisoning and the toxin involved).

Facial eczema disease of cattle and sheep is caused by toxic spores of the fungus *Pithomyces chartarum* that infests lowland pastures in the North Island and northern South Island. The fungus is indigenous to New Zealand, but is also found in Australia and South Africa. Sporidesmin, the toxin that leaches from the spores, was isolated in 1959 by R.L.M. Synge and E.P. White,[54] and there is now a considerable literature on it.

Karaka (*Corynocarpus laevigatus*) is mentioned in the sections on carbohydrates and insecticides in Chapter 10, Chemical Products from Native Plants. Apparently the fruit may cause paralysis of the hind limbs and cessation of lactation in cows;[53] the nectar is poisonous to bees, and the kernels, if not soaked and cooked, are poisonous to humans.

The leaves of ngaio (*Myoporum laetum*) are poisonous to cattle, and ngaione has been isolated from them. Ngaione is toxic, causing liver damage similar to that seen in facial eczema of sheep.

The most dangerous of the native plants is probably ongaonga (*Urtica ferox*) which is also known as 'tree nettle'. As indicated earlier its stinging hairs have been known to kill a man, horses, and dogs. The nature of the toxic principle, isolated by F.N. Fastier and G.F. Laws,[50] is unknown.

Pinātoro (*Pimelea prostrata*) is also a common cause of poisoning of horses and cattle. Its active constituent prostratin was described by the late Professor B.F. Cain as one of the most toxic he had encountered in his survey of the native flora.

Tutu (*Coriaria* species) was responsible for deaths among the first stock brought to New Zealand, and the shoots poison some cattle each year. Its discovery was recorded by George Forster[56] (see Chapter 2), and its role as a food plant is discussed in Chapter 8.

Of the native pasture plants, waoriki (*Ranunculus amphitrichus*, formerly *R. rivularis*) causes losses of cattle; the active principle is a lactone protoanemonin which causes irritation of tissues and is released from the glycoside ranunculin by enzymic hydrolysis. Other native plants which are poisonous have been dealt with elsewhere, for example *Solanum* species, *Sophora* species, *Laurelia novae-zelandiae* and *Pseudowintera axillaris*, as medicinal plants.

There is a possibility that some of the native poisonous plants may have a role to play as substitutes, if ever our drug supply is threatened. Their active principles include alkaloids (e.g. *Solanum* species), cyanogenic glycosides (e.g. *Alectryon excelsus*), terpenoids (e.g. *Coriaria* species, *Myoporum laetum*, *Pimelea prostrata*), nitrates, oxalates and nitro-compounds (e.g. *Corynocarpus laevigatus*.[1]

1. Brooker, S.G., Cambie, R.C., Cooper, R.C. *New Zealand Medicinal Plants* (3rd ed.) Auckland (Heinemann) 1987 268p.
2. Gluckman, L. *Medical History of New Zealand Prior to 1860* Auckland (Published by the author) 1976 271p.
3. Webb, L.J. *Guide to Medicinal and Poisonous Plants of Queensland* C.S.I.R.O. Bull. 232, Melbourne (Government Printer) 1948 202p.
4. Schiffer, A.P., Kovacs, A. (1959) 183 *Nature* 988.

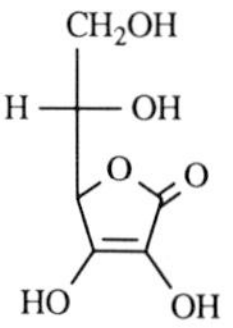

(1) ascorbic acid
(vitamin C)

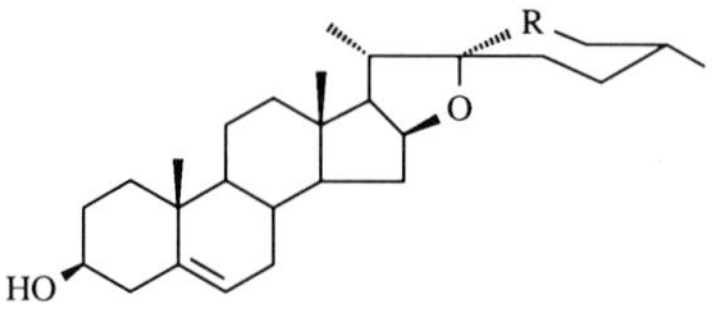

(2 R = NH) solasodine
(3 R = O) diosgenin

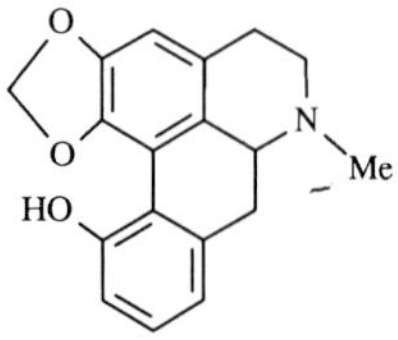

(4) pukateine

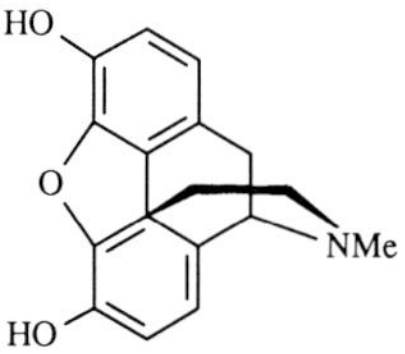

(5) morphine

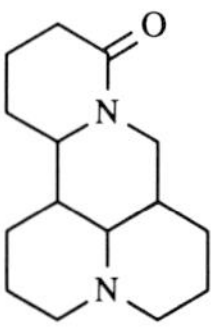

(6) matrine

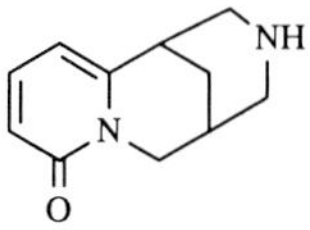

(7) cytisine

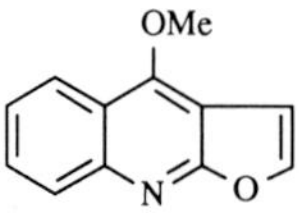

(8) dictamnine

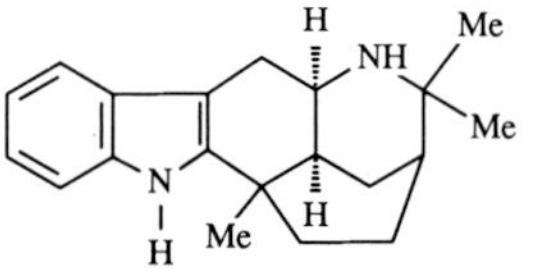

(9) aristoteline

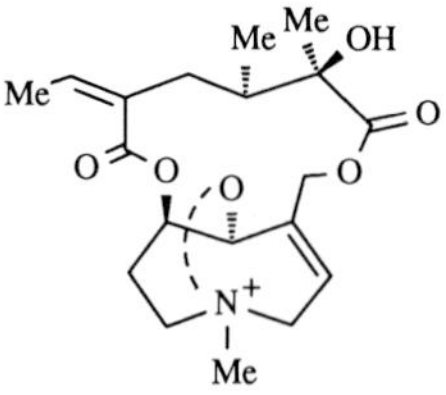

(10) senkirkine

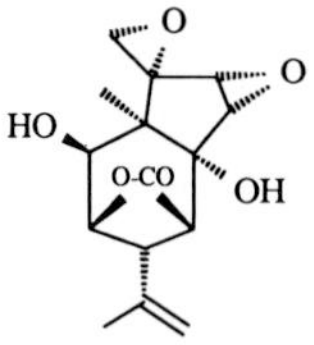

(11) tutin

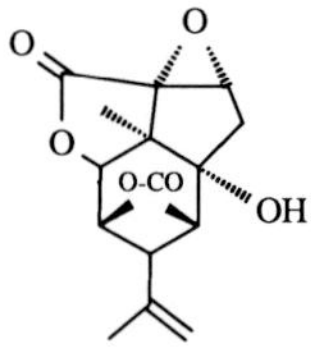

(12) picrotoxinin

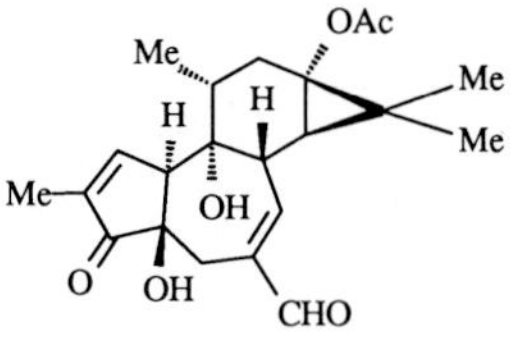

(13) prostratin

5. Brown, T.E., Eyster, H.C. (1955) 55 *Ohio J. Sci.* 257 (Chem. Abs. 1956, 50, 5009).
6. Okuyama, E., Yamazaki, M. (1983) 31 *Chem. Pharm. Bull Jpn.* 2209.
7. Wharton, W.J.L. (ed.) *Captain Cook's Journal during his first Voyage around the World* London (Eliot Stock) 1893 400p.
8. McNab, R. (ed.) *Historical Records of New Zealand*, Wellington (Government Printer) 1914 2 vols (reprinted 1973). Extracts from the Journal of P. de L'Horne, 1769-1770.
9. Best, E. (1905) 13 *J. Polynes. Soc.* 213.
10. Best, E. (1906) 14 *J. Polynes. Soc.* 1.

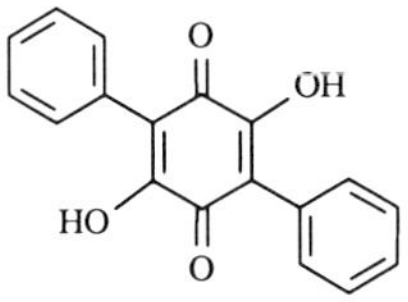

(14) polyporic acid

CH₃CO₂CH₂CH₂NMe₃OH

(15) acetylcholine

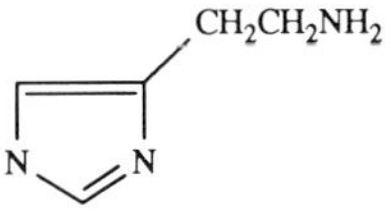

(16) histamine

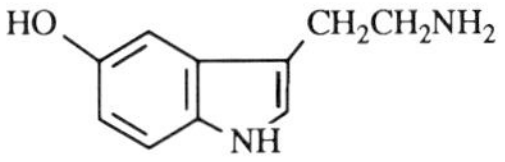

(17) 5-hydroxytryptamine

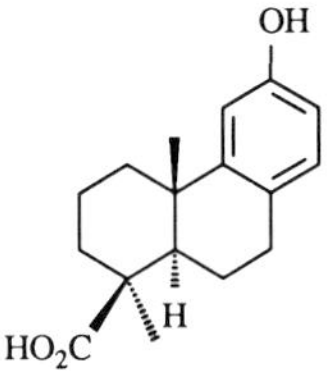

(18) podocarpic acid

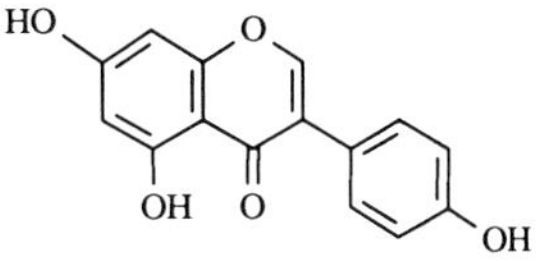

(19) genistein

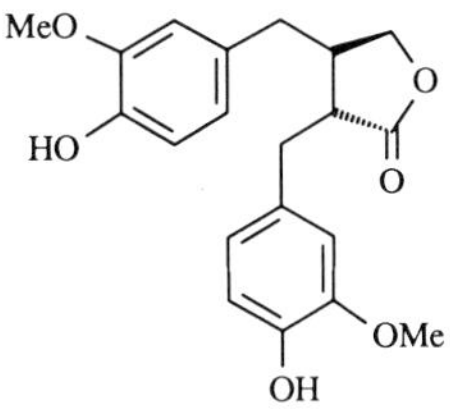

(20) matairesinol

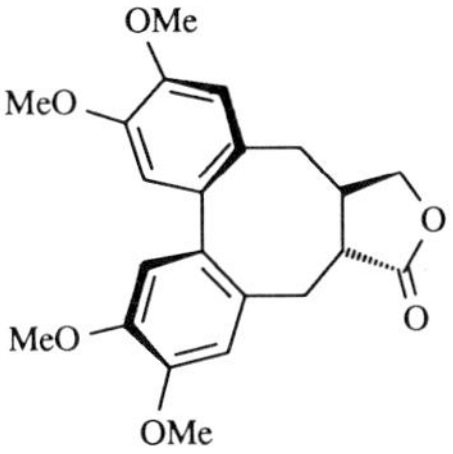

(21) an isostegane

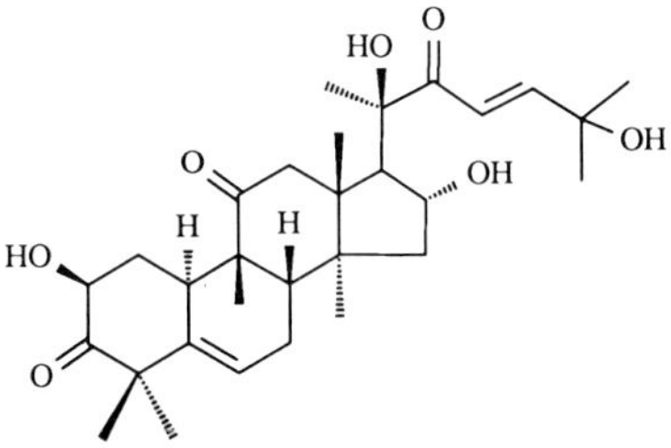

22 cucurbitacin D

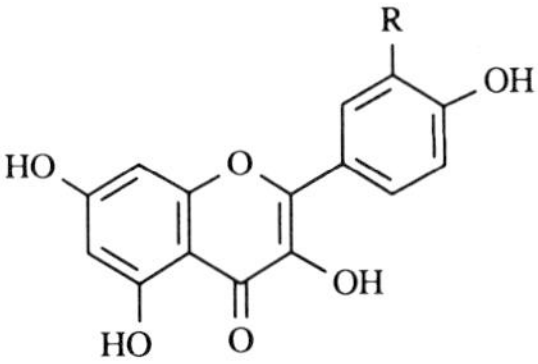

(23) a R = H   quercetin
     b R = OH kaempferol

11. Best, E. (1908) 40 *Trans. N.Z. Inst.* 185.
12. Anon. *The New Zealanders* London (Charles Knight) 1830 424p.
13. Cruise, R.A. *Journal of a ten months residence in New Zealand* (2nd ed.) London (Longman, Hurst) 1824 327p. (reprinted by Capper Press, Christchurch 1974).
14. Fulton, R. *Medical Practice in Otago and Southland in the Early Days* Dunedin (Otago Daily Times & Witness) 1922 304p.
15. Colenso, W. (1869 2nd ed. 1875) 1 *Trans. N.Z. Inst.* 233.
16. Williams, W. *Journals*, 1825-1855, typescript copies in Library, Auckland Institute and Museum.

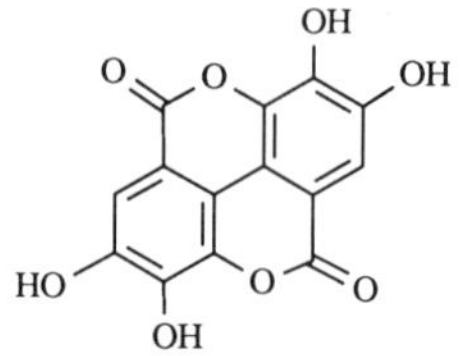

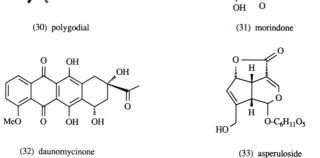

(24) ellagic acid

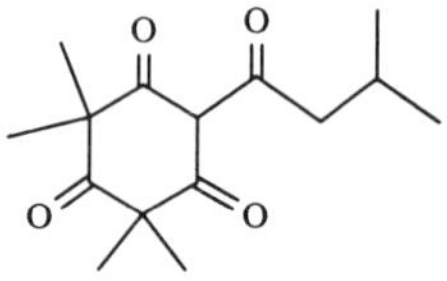

(25) leptospermone

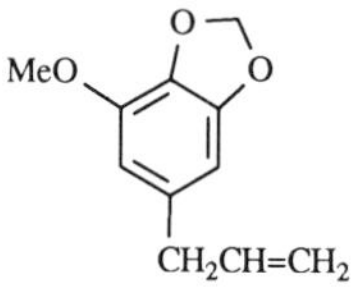

(26) myristicin

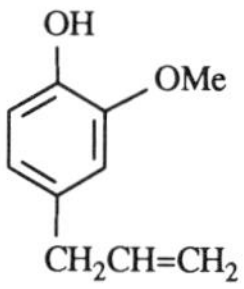

(27) eugenol

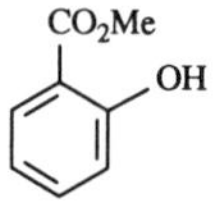

(28) methyl salicylate

$$CH_2=CHCH(OH)(C≡C)_2CH(OH)CH=CHC_7H_{15}$$

(29) falcarindiol

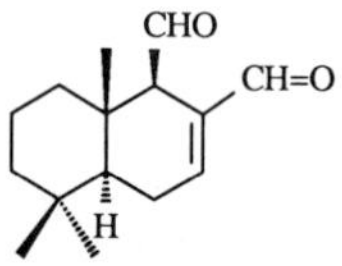

(30) polygodial

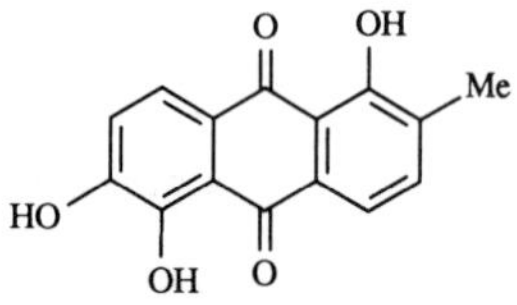

(31) morindone

(32) daunomycinone

(33) asperuloside

17. Taylor, R. *Journals*, 1833-1873, typescript copies in Library, Auckland Institute and Museum.

18. Baker, C. *Journals*, 1827-1867, typescript copies in Library, Auckland Institute and Muscum.

19. Chapman, T. *Letters and Journals*, 1830-1869, typescript copies in Library, Auckland Institute and Museum.

20. Puckey, W.G. *Letters and Journals*, 1831-1868, typescript copies in Library, Auckland Institute and Museum.
21. Neil, J.F. *The New Zealand Family Herb Doctor* Dunedin (Mills, Dick & Co.) 1889 512p 2nd ed. 1891 524p (reprinted by Capper Press, Christchurch 1976).
22. Goldie, W.H. (1905) 37 *Trans. N.Z. Inst.* 1.
23. Martin-Smith, M. Pers. commun. to the late Prof. L.H. Briggs 1968.
24. Hunt, P. (daughter of Rev. E. Ward) Pers. commun. 1981.
25. Rafter, P.M. *Never Let Go! The Remarkable Story of Mother Aubert* Wellington (Reed) 1972.
26. Adams, O.L.G. (1945) *Maori Medicinal Plants* Auckland Bot. Soc. Bull. No. 2.
27. Macdonald, C. *Medicines of the Maori* Auckland (Collins) 1973 142p.
28. Palmer, G.B. (1954) 63 *J. Polynes. Soc.* 147.
29. Cain, B.F., Scannell, S., Cambie, R.C. (1961) 4 *N.Z.J. Sci.* 3.
30. Cambie, R.C., Cain, B.F., La Roche, S. (1961) 4 *N.Z.J. Sci.* 604, 707, 731.
31. Cambie, R.C. (1971) 35 *Chem. in N.Z.* 69.
32. Cambie, R.C. (1976) 6 *J. Roy. Soc. N.Z.* 307.
33. Fogg, W.S. (1935) 54 *J. Pharm. Exp. Therap.* 167.
34. Burton, J.F., Cain, B.F., (1959) 184 *Nature* (London) 1326.
35. Brandt, C.W., Ross, D.J. (1948) 161 *Nature* (London) 892.
36. Bible, R.H. (1956) United States Patents 2750270, 2750373, 2753357, 2759014; (1957) ibid. 2786431, 2796431; (1958) ibid. 2851485, 2854474; (1960) 2971008.
37. Bible, R.H., Hoehn, N.M. (1956) United States patents 2705725; (1959) ibid. 2889371.
38. Cambie, R.C., Clark, G.R., Craw, P.A., Rutledge, P.S., Woodgate, P.D. (1984) 37 *Aust. J. Chem.* 1775.
39. Harvey, H., Waring, J. Pers. commun. 1985
40. Kupchan, S.M., Meshulam, H., Sneden, A.T. (1978) 17 *Phytochem.* 767.
41. Anon. (1989) *New Zealand Herald*, 12 January.
42. Jones, M. (1989) *New Zealand Herald*, 15 September; Russell, K.M., Molan, P.C., Wilkins, A.L., Holland, P.T. (1990) 38 *J. Agric. Food Chem.* 10.
43. Taylor, R. (1848) *New Zealand Journal* (London) No. 217, 25 March, p. 68; extracted from the *N.Z. Church Almanack* for 1847.
44. Muir, A.D., Cole, A.L.J., Walker, J.R.L. (1982) 44 *Planta Med.* 129.
45. Muir, A.D., Walker, J.R.L. (1979) 43 *Chem. in N.Z.* 94.
46. McCallion, R.F., Cole, A.L.J., Walker, J.R.L., Blunt, J.W., Munro, M.H.G. (1982) 44 *Planta Med.* 134.
47. Anon. (1984) *Chem. in Britain* 598.
48. Boddy, I.K., Boniface, P.J., Cambie, R.C., Craw, P.A., Zhen-Dong, H., Larsen, D.S., McDonald, H., Rutledge, P.S., Woodgate, P.D. (1984) 37 *Aust. J. Chem.* 1511.
49. Cambie, R.C. and co-workers, University of Auckland, unpublished work, 1984.
50. Fastier, F.N., Laws, G.F. (1975) 6 *Search* 117.
51. Calder, V.L., Cole, A.L.J., Walker, J.R.L. (1986) 16 *J. Roy. Soc. N.Z.* 169.
52. Molloy, B. (1990) 130 *Botany Division Newsletter* 3.
53. Connor, H.E. *The Poisonous Plants in New Zealand* Wellington (Govt. Printer), 2nd ed. 1977 247p.
54. Evans, I.A., Osman, M.A. (1974) 250 *Nature* (London) 348.
55. Synge, R.L.M., White, E.P. (1959) *Chem. and Ind.* 1 546.
56. Forster, G. *De Plantis Esculentis Insularum Oceani Australis Commentatio Botanica* Berolini (Haude and Spener) 1786, 80p.

10

Chemical Products from Native Plants

10.1 Literature of New Zealand Natural Products

The chemical examination of the endemic flora has for many years been a productive avenue for research in the universities and government research institutions of New Zealand. Initial work was carried out by W. Skey from 1865 to 1900 at the Colonial Laboratory in Wellington, and this was followed by a fruitful partnership between T.H. Easterfield, first Professor of Chemistry at Victoria College, Wellington, and B.C. Aston, first chemist in the Department of Agriculture, who was also a competent botanist. One of the most productive schools of research to continue in this area was that of L.H. Briggs at the University of Auckland. Briggs was associated with the late Dr B.F. Cain, who carried out an unpublished survey of a large number of native plants in search of compounds with anticancer activity. Many projects with commercial potential have been, and continue to be, investigated at the Chemistry Division of DSIR at Petone, the Crop Research Division of DSIR at Lincoln, and at the Applied Biochemistry Division of DSIR at Palmerston North.

Partly because of the small numerical size of the flora, its chemical literature has been well surveyed. Early reviews were published by Gardner,[1,2] Murray,[3,4] and Briggs;[5] and in 1963 and 1966 Brooker, Cambie, and co-workers[6,7] published a phytochemical register, containing references to recorded isolation of constituents. This survey has been updated at approximately ten year intervals.[8,9] In addition, a phytochemical survey of many New Zealand plants for alkaloids and other classes of compounds (saponins, leucoanthocyanidins) has been carried out by Cain and Cambie[10-12] and by Bailey[13] (monosaccharides) and Russell[14] (ecdysones). A complementary survey of New Zealand plants for antibiotic activity against a range of bacteria, yeasts, and fungi, by Calder, Cole and Walker has also been published.[15] Brooker, Cambie and Cooper[16] have published a book entitled *New Zealand Medicinal Plants* in which the medicinal uses of the native flora

William Skey, Colonial Analyst, Wellington

are recorded, along with the constituents which might be responsible for their value. More recently the latter three authors have published a brief paper which surveys the actual and potential economic uses of our native plants.[17] Constituents of the indigenous native flora have been reviewed by Cambie with a view to highlighting those of actual or potential economic value.[18]

While the milling of native timbers such as kauri, rimu, tōtara, and matai provided rich returns to sawmillers, only limited attention was paid by early settlers in New Zealand to the utilization of sawdust and wood-waste. Processes such as (i) the dry distillation of wood to produce wood spirit (methyl alcohol), acetone, acetic acid, and charcoal, (ii) chemical treatment to produce cellulose, vinegar, alcohol, sugar, and oxalic acid, and (iii) miscellaneous processes such as the production of particle board, were for the most part neglected, primarily because of the sparse population and the lack of competition for such products. Nevertheless, over the years some economically important work has been carried out, primarily in the Chemistry Division of the DSIR and within certain New Zealand universities, which has been aimed at utilizing the plant products of our flora. Important classes of compounds with economic potential include dyestuffs, tannins, kauri gum, insecticides, saponins, carbohydrates, resin acids and resinols, and perfumes. In this chapter we summarize the more important aspects of this work.

10.2 Dyestuffs (Plate 15)

One of the things which impressed the early European explorers in this country was the striking jet-black colour of plant fibres dyed by the Māori. Although red-brown colours and occasionally a deep golden colour were also observed, the fastness and intensity of the black dyes were a source of astonishment. Dyewoods were part of the experimental cargo of the *Bolina* in 1844, referred to in Chapter 2. The use of the dyes by the Māori was described[20-23] by Aston in a series of papers published between 1917 and 1919. According to Aston only three plants were commonly employed for the production of fibres (but see Chapter 5): the well-known New Zealand flax (*Phormium tenax*), the closely related mountain flax (*Phormium cookianum*, formerly *P. colensoi*), and the cabbage tree (*Cordyline australis*). The fibres were dyed in the unmanufactured state, and the latter plant was almost invariably dyed black.[24] The flax would not take a dye unless the skin or dermis was broken and bruised, enabling the Māori to produce a variety of design effects of which tukutuku work provides some very attractive examples.

The chief source of the black dye was the bark of hīnau (*Elaeocarpus dentatus*). As early as 1817 Nicholas[25] related how he had observed near a Māori dwelling a vessel in the shape of a flat-bottomed bowl (kumete) in which bark of the 'enou' was being steeped prior to dyeing flax. Often the fibre was placed in black mud of a kahikatea (*Dacrycarpus dacrydioides*) swamp for a lengthy period, the iron salts in the mud acting as a mordant. Black dyes were also obtained from the pōkākā (*Elaeocarpus hookerianus*) and to a lesser extent from beech trees (*Nothofagus fusca* and *N. menziesii*), kāmahi (*Weinmannia racemosa*), and tawhero (*W. silvicola*). Blue-black colours were obtained from tutu (*Coriaria arborea*), and makomako (*Aristotelia serrata*). A red-coloured dye was obtained from tānekaha (*Phyllocladus trichomanoides*) or toatoa (*P. asplenifolius* var. *alpinus*) while yellow colours came from pūriri (*Vitex lucens*) (Plate 13) or from *Coprosma* species.

In the late 1800s a considerable amount of tānekaha bark was exported to Germany by Messrs. Krull and Co.;[20-22] it contains a high percentage of catechuic and tannic acids, and gives reddish and pink colours with zinc, tin, and aluminium mordants, but the colours are too drab to be of much value.

A.G. Perkin, working in the Clothworkers' Research Laboratory of the Yorkshire College Dyeing Department, examined the dyeing properties of the wood of pūriri in 1898.[26] The most striking property of the dyestuff was the strong yellow tones produced with chromium and aluminium mordants. Perkin reported that a well-known Leeds firm, which had examined a sample, considered it should have a limited sale at four pounds per ton, a value somewhat below that of old fustic (from *Morus tinctoria* now *Chlorophora tinctoria*) one of the natural dyestuffs then employed in industry.[20-23] The principal colouring matter of *Vitex lucens* is vitexin (see fig. formula 1), a flavonoid C-glucoside, whose structure was determined subsequently.[8]

Following a resolution of the Farmers' Union at their Dominion conference in 1916 in which they expressed a desire to use New Zealand vegetable dyes for homespun woollen goods and the dyeing of military material,[27] Aston evaluated New Zealand *Coprosma* species for their dyeing properties. Of the more abundant species, *C. australis* (then known as *C. grandifolia*) and *C. areolata* provided the best results, yielding dyes by extraction with dilute sodium hydroxide. With chrome mordants, these gave purplish or maroon fast dyes with wool, and tangerine to dark shades with stannous mordants and tartar emetic. In his papers,[23] including one which was illustrated by pieces of wool coloured by the dyes, Aston detailed the extraction processes and the effect of alkali on a number of *Coprosma* species. *Coprosma areolata* was of special importance, since it yielded a direct or substantive dye which could be applied without the use of a mordant.

Given the interest in the establishment of home spinneries, Aston hoped that New Zealand homespuns made from wool by soldiers returning from World War I, and coloured by dyestuffs obtained from local plants, might 'become as fashionable and command as good a price as the celebrated Harris tweeds of fragrant memory'.[20-23] However, this was not to be. Although home spinning has continued to be a popular pastime of many people, and while numerous enthusiasts have successfully dyed their garments with extracts of native plants, little of commercial importance has arisen in this area.

The task of identifying the colouring matters in *Coprosma* species as anthraquinones was taken up by L.H. Briggs at the University of Auckland. In a series of papers published from 1948 to 1963, he and his students examined and identified the anthraquinones in most of the common species.[8] R.D. Wilson[28,29] has recently used the flavonoid glycosides of *Coprosma* species as a means for identifying hybridization in the genus.

10.3 Tannins

One of the first forest by-products to be utilized by early settlers in New Zealand was the bark from certain native trees for the tanning of leather. As early as 1839, Colenso had observed hides tanned with native barks at Ngunguru, near Whangarei.[20-23] By 1843 a tannery was operating at Russell and two years later there were tanneries at Auckland and Nelson. The Nelson tannery, built by G.W.W. Lightband at the Matai River, had vats of pukatea (*Laurelia novae-zelandiae*), and used 'hīnau and black birch' for tanning. Hīnau is *Elaeocarpus dentatus*, and black birch is a popular name for species of *Nothofagus, Pittosporum tenuifolium*, and *Weinmannia racemosa*. The subsequent development of tanneries has been described by G.G. Thornton.[30] Most of the plants which had been found to be rich in dyestuffs were also found to be rich in tannins. Chief among these was tānekaha (*Phyllocladus trichomanoides*) which, with its large tannin content, was highly prized. According to Kirk[31] (in 1889), it possessed special value as an organic mordant in the preparation of basils for kid gloves.[20-22] The bark was exported for many years, some fifty-five tons leaving the country in 1931 for a return of $1380.

Colenso[32] recorded that the bark of tōwai (*Weinmannia racemosa*) was used extensively for tanning and had been exported, while Kirk[31] wrote that the bark 'is of great value for tanning as it contains 13 per cent of tannin'. Kirk also commented that the allied tawhero (*W. silvicola*)

> has long been famous for its bark, which contains 10 to 13 per cent of tannin, and was formerly used to a large extent in the Auckland tanneries: the supply, however, was obtained in a most wasteful manner, the bark being peeled as high as a man could reach, the branches and upper portion of the trunk being left untouched, so that the supply in the vicinity of settlements soon became exhausted.

It is of interest that one hundred years later *Weinmannia* is extraordinarily rare in the Waitakere Ranges and about Auckland city;[33] the ravages of the tanners may be an explanation of this scarcity.

Skey analysed the barks of a number of indigenous plants in the Colonial Laboratory during the late 1800s, and later Aston summarized the tanning agents which could be found in New Zealand plants.[20-22] Attempts were made to utilize some of these materials, but in general the rate of tanning was slow compared with Australian or Tasmanian wattle. By 1916, the Woolston Tanneries Ltd of Christchurch had come to the conclusion that only kāmahi (tōwai: *Weinmannia racemosa*) was worthy of further experimentation. However, Aston noted that excellent leather had been made in the Nelson area using beech (*Nothofagus*) bark, and suggested that practical tanning tests using hides were necessary in order to determine the value of native barks as tanning agents. However, by 1918 a shortage of imported tanning materials was over and the tests were never made.[16]

Tannins are water-soluble polymers which may be conveniently divided into two major subdivisions: hydrolysable tannins and condensed tannins. Hydrolysable tannins break down on acid hydrolysis to either gallic acid and glucose (gallotannins) or to ellagic acid and glucose (ellagitannins). Little is known of their detailed structures but it has been suggested that they are polyesters formed by self-condensation of gallic acid or of its oxidative coupling products with occasional glucose units attached to the main polymer chain by way of ether or ester linkages. Under the same conditions of acid hydrolysis, condensed tannins precipitate insoluble substances known as 'phlobaphenes' or 'tanners' red'. Owing to their resistance to chemical degradation even less is known about them except that they may be polymeric flavans, related to flavonoids. Hydrolysable and condensed tannins may often occur together. Although natural tannins may differ greatly in chemical constitution, all will precipitate proteins from solution or will react with them. It is this property which is responsible for their astringent qualities.

10.4 Kauri Gum

One of the best known New Zealand natural products to have been utilized commercially is kauri gum. The gum is mainly a subfossil resin, but it oozes freely from the kauri (*Agathis australis*) whenever an incision is made. It was first exported to London as an experimental consignment in the 1830s, and between 1845 and 1900 production increased steadily as gumfields were opened in Northland and Coromandel, reaching a peak in 1899, when approximately 11,000 tonnes were produced. At the same time the price rose gradually to reach a peak of eighty-three pounds per tonne in 1920. Although some 7000 tonnes were exported in 1924, mainly to American traders, production decreased and dwindled to small amounts over the next ten years and by 1950 the industry was virtually defunct. Contrary to popular belief, it was the supply and cost of the gum rather than the superior quality of synthetic alternatives which led to the decline in the industry.

Evaporation of volatile components, retention of others as plasticizers, and a moderate degree of polymerization in air, all cause kauri gum to have useful coating properties. The gum was originally used by varnish makers in England and the US because it united with linseed oil faster and at a lower temperature than any other resin of comparable nature. Later, it was used mainly for the production of linoleum. Attempts to bleed living trees to produce gum result in the eventual death of the tree; this method is therefore strongly deprecated. It is interesting to note that the 'kauri-butanol value' is still a standard in the varnish industry in which solvents are rated by the amounts which can be added to a standard solution of kauri gum in butanol without causing turbidity.[35]

Scientists at the DSIR contributed much to the investigation and exploitation of kauri gum. According to William Donovan, J.S. Maclaurin, Colonial and later Dominion Analyst, devoted much time to research on kauri gum, but little if any of his work was ever published. Donovan, who was Maclaurin's successor, developed solvents for kauri gum and for the resin in buried kauri timber. He also patented a process for the purification of fine swamp-gum based on differential flotation in a salt solution of defined density before and after application of a vacuum. Apparently this was used successfully on a commercial scale in Auckland.[27] During the 1930s Depression, Brandt and Hosking set up a pilot plant for refining the poorer 'chalk or chip' grades of kauri gum from 'pickings' of the gum fields based on solvent extraction, while work was done both in Auckland and Wellington on the removal of colour from the darker grades of kauri gum; however, there was 'no satisfactory commercial outcome'.[27]

The chemistry of modern and fossil kauri resins has been reviewed by Thomas.[36] Patterns of kauri constituents are recognizable in resins up to sixty million years old, this resistance to biodegradation presumably being related to the hard mechanical barrier set up when the resin flows and hardens. Early work on the chemical constituents was carried out by J.R. Hosking[37] who, in 1929, isolated the diterpenoid agathic acid (2) in addition to the common terpenes α-pinene (3) and limonene (4). However, it was not until 1964 and 1966 that L.J. Gough[38] in London and B.R. Thomas[39] at the DSIR in Petone investigated the bled resin and isolated a series of diterpenoid acids, *viz.* abietic, *cis*- and *trans*-communic, sandaracopimaric, and agathic acids, and their corresponding alcohols.

Kauri gum is apparently still plentiful in Northland, and in recent years there has been a renewed interest in the possibility of its commercial utilization. In 1979[40] a firm, Kauri Deposit Surveys Ltd (later Kaurex Corporation Ltd), applied for a licence to extract waxes and resins from the peat swamp at Kaimaumau, adjoining Rangaunu Harbour in Northland. After four years of planning, environmental impact reports, and government scrutiny, approval of the scheme was granted in late 1983.[41] The proposal was to mine the peat and extract the dried material with hot butyl acetate to give a product which was estimated from pilot plant work to be in the order of 8-9% of the dry weight of the peat. The crude product would be further refined to give a bituminous resin whose commercial value would be enhanced by successive fractionations into resin and wax materials. The price of the final product on the international market was estimated to be more than $4000 a tonne, and indicated reserves of 530,000 tonnes were estimated to provide for a $3 million a year industry. The products would have use in surface coatings, linoleum, rubber compounding, adhesives, paper coatings, and ink.[42] Original plans included the use of the processed peat as a fuel for the plant's boilers. However, the high sand content of the peat clogged the boilers, making it necessary to substitute coal and electricity as the fuel, adding unexpected operating costs to the project. These and other design faults resulted in the firm going into voluntary receivership[43,44] in early 1988, and the plant was dismantled.[45]

There are a number of publications on the kauri gum industry;[46-48] the reader is referred to these for further information. Kauri gum remains a prominent material in souvenir shops.

10.5 Insecticides

From the oil of the mānuka (*Leptospermum scoparium*) L.H. Briggs and co-workers isolated an unusual constituent which they named leptospermone (5). This was shown to have anthelmintic properties.[49,50] Its structure was similar to that of synthetic 1,3-indandiones such as valone (6), which has high insecticidal properties.[58] Tests on the insecticidal properties of leptospermone showed it to have a good synergistic effect, but it was inferior to valone.[5]

3-Nitropropanoyl-D-glucopyranoses such as karakin, caronarian, and cibarian found in karaka (*Corynocarpus laevigatus*) are toxic to the grass grub (the larva of *Costelytra zealandica*). These compounds are also found in the forage legumes *Lotus pedunculatus* and *Coronilla varia* (crown vetch)[52] and could have potential as insecticides.

A recent development has been the isolation from plant sources of polar steroids, which show a marked moulting hormone activity when injected into insects. These hormones initiate the pupal and larval stages through which an insect grows to maturity. The uptake of excessive amounts of these compounds causes a lethal change in the normal development of the insect. These compounds, called 'ecdysones', were originally isolated from insects and crayfish, in which they occur in extremely small amounts. In 1970 G.B. Russell of the Applied Biochemistry Division, DSIR, Palmerston North, carried out a survey of New Zealand gymnosperms and showed that the barks of a high proportion of them show insect moulting-hormone activity.[53] The compounds are also present in twenty-four of sixty-four New Zealand ferns examined,[54] the total number of species and subspecies of ferns and fern-allies here being about 190.

In the US, attempts have been made to utilize ecdysones as commercial insecticides. Thus it is somewhat surprising that there has been little or no interest in the commercial possibilities of ecdysones in this country in view of our rich source of such compounds, our problems with insect control, and the current use of insecticides with harmful effects on other biological material. Compounds responsible for the earlier metamorphoses of insects are called juvenile hormones. These compounds have linear structures based on farnesol, and have a similar insecticidal potential to that of ecdysones. Analogues of juvenile hormones containing an aromatic ring, for example juvadecene (7), have recently been identified as active constituents of the roots of kawakawa (*Macropiper excelsum*).[55] Juvadecene induces supernumary metamorphosis when applied to last-instar nymphs of the milkweed bug at a concentration of 30microg, and is quite toxic at higher doses.[56] The chemistry of ecdysones is discussed in Chapter 12.

10.6 Saponins

Saponins are compounds which consist of either steroids or triterpenoids joined to sugar moieties. On acid hydrolysis they give sapogenins which find use as starting materials for the production of steroidal hormones. In a comprehensive survey, Marker and co-workers[57] examined over 300 species from the Central American region for saponins. Many of the species belong to the family Agavaceae which, in New Zealand, was represented by the familiar cabbage tree (*Cordyline australis*), four other species of the same genus, New

Zealand flax (*Phormium tenax*) and mountain flax (*P. cookianum*). However, lately *Cordyline* and *Phormium* have been separated from the Agavaceae and placed in smaller plant families; *Cordyline* is now in the cabbage tree family, the Asphodelaceae, and *Phormium* is in the Phormiaceae. The leaves of New Zealand *Cordyline* species have been investigated recently.[58] All contain 1,3-dihydroxysapogenins, but it is doubtful if they are present in sufficient quantity for commercial exploitation.

Other plants of the New Zealand flora have been surveyed for the presence of saponins[10-12] and some, such as *Pittosporum* species and *Dodonaea viscosa*, may be worthy of further investigation. Leaves of kūmarahou (*Pomaderris kumeraho*), also called 'gumdigger's soap', contain appreciable amounts of saponins, but these have not been investigated in detail.

10.7 Carbohydrates

Simple monosaccharides such as arabinose, xylose, rhamnose, glucose, galactose, and fructose have been isolated or shown to be present in a number of New Zealand plants[13] while disaccharides such as sucrose occur in *Coprosma* and *Sophora* species as well as others.[8] The concentration of fructose in the roots and stems of the cabbage tree (*Cordyline australis*) is reputed to be high enough to warrant development of commercial plantings. The yield compares favourably with that from sugar beet. Fructose is almost twice as sweet as sucrose syrup and thus is an excellent sweetening agent, being used widely in food processing[59] (see Chapter 8).

Little work has been carried out on polysaccharides, and the gums present in many New Zealand plants, for example *Meryta sinclairii*, have been almost entirely neglected in this country, probably because of the considerable difficulties encountered in working with these substances. The gum from *Phormium tenax* contains a hemicellulose composed of an aldotrionic acid with two nine-unit xylose chains attached to its terminal xylose residue.[60,61] More recently a mucilage or gum has been isolated from the leaves of the whau (*Entelea arborescens*) which could have potential commercial use as an additive in the food industry or for other industrial applications. Chemically the mucilage is similar to imported gums and is composed of residues of L-arabinose, D-galacturonic acid, L-rhamnose, and D-galactose with trace amounts of D-glucose and D-xylose.[62]

Of interest is the fact that kernels of berries of karaka (*Corynocarpus laevigatus*) were soaked, cooked, and eaten by the Māori. The kernels contain a poisonous glucoside, karakin ($C_{20}H_{30}N_4O_{19}$), which is hydrolysed during the soaking and cooking processes to give hiptagenic acid (α-nitropropionic acid) and glucose.[63,64] More recently, other related glucosides of similar structure have been isolated from the same source.[65] It is probable that the glucose provided the culinary attraction of the berries (see Chapter 8).

The four abundant New Zealand seaweeds of the Gelidiaceae, *Pterocladia lucida*, *P. pinnata*, *Gelidium caulacantheum*, and *G. pusillum* contain polysaccharides based on the classical agar structure.[66]

10.8 Alkaloids (other than *Solanum*)

As indicated in Chapter 9, it is usually alkaloid-bearing species which attract attention for their medicinal uses, and indeed many of our New Zealand medicinal plants are those which contain alkaloids. New Zealand species have been surveyed for alkaloids[8], but this country contains few members of those plant families (Apocynaceae and Rutaceae are two examples) which are traditionally recognized as alkaloid-bearing.

The occurrence of alkaloids with medicinal potential in the pukatea (*Laurelia novae-zelandiae*), and kōwhai (*Sophora microphylla*) and related species of the Fabaceae is dealt with in Chapter 9. Other species which have been found to contain alkaloids of note are mairehau (*Phebalium nudum*),[67] Kirk's daisy (*Senecio kirkii*), now *Brachyglottis kirkii*),[68] makomako (*Aristotelia serrata*)[69] and kaihua (*Parsonsia heterophylla*),[70] their families being the Rutaceae, Asteraceae, Elaeocarpaceae, and Apocynaceae respectively.

Kirk's daisy (*Brachyglottis kirkii*) **Asteraceae**
Painting: F. Osborne

kaihua (*Parsonsia heterophylla*) **Apocynaceae**
Engraved from a 1769-70 painting by Sydney Parkinson

10.9 Oils and Fats

Except for an extensive examination by I.M. Morice[71] of fatty acids of the seed oils from New Zealand monocotyledons such as members of the Agavaceae, relatively little work has been carried out on the non-volatile oils and fats from native species. S.G. Brooker has investigated the seed oil of kōhia, the native passionfruit (*Tetrapathaea tetrandra*)[72] and the oil of the scarlet aril and seeds of tītoki (*Alectryon excelsus*);[73] Morice has studied the seed and fruit coat acids of hīnau drupes (*Elaeocarpus dentatus*);[74] while Borges del Castillo and co-workers have investigated the alkanes of leaf waxes of most New Zealand gymnosperms.[75] Body has reported on the fatty acid composition of the oil from the kernels of karaka fruits (*Corynocarpus laevigatus*).[76] The oil of kōhia is a rich source of the essential fatty acid, linoleic acid.

karaka (*Corynocarpus laevigatus*) **Corynocarpaceae**
Engraved from a 1769-70 painting by Sydney Parkinson

Morice[77] reported that up to 25% of the total seed oil of New Zealand *Astelia* species is γ-linolenic acid (GLA) which compares favourably with *c.* 9% in the seeds of evening primrose (*Oenothera biennis*) which is being actively promoted as a potential medicament in a range of medical conditions varying from pre-menstrual syndrome to multiple sclerosis. GLA is an essential fatty acid required by the body, as a metabolic precursor for prostaglandins. Such GLA-rich oils are in growing demand as nutritional supplements. It is claimed that small doses of linolenic acid can overcome a deficiency caused by processing vegetable oils, some medical conditions such as diabetes, or the process of ageing.[74,78] Two of the species (*A. trinervia* and *A. solandri*) are comparatively common on the West Coast and it has been suggested that their cultivation for commercial purposes might merit attention.

There has been an extensive investigation of the oils and fats present in some thousands of domestic and foreign plants by workers at the Northern Regional Research Center of the US Department of Agriculture at Peoria, Illinois.[79] In the light of their results from tests on other plants of the Sapindaceae, Brooker and Eyres[80] re-examined the oil of the tītoki and found it to contain about 40% of compounds with a cyanogen group, for example (8). It is likely that other interesting compounds could be found in the non-volatile oils of our native flora, especially with the more sophisticated techniques which are now available.

10.10 Resin Acids and Resinols

One of the most notable of the resin acids obtained from New Zealand species is agathic acid (2) from the kauri (*Agathis australis*). It is one of the principal constituents of kauri gum (see above) and has been the subject of a number of chemical investigations[37] culminating in the elucidation of its structure and in the investigation of the range of constituents in both fresh and fossil resins.[38,39] The economic use of kauri gum is also dealt with earlier in this chapter.

The oleo-resins or the woods of most New Zealand members of the family Podocarpaceae have been examined, and details can be found in the phytochemical register of the extractives from New Zealand plants.[8,9] Of particular interest from an economic point of view are the resins of rimu (*Dacrydium cupressinum*), kahikatea (*Dacrycarpus dacrydioides*), mataī (*Prumnopitys taxifolia*), N.Z. pink pine (*Halocarpus biformis*, formerly *Dacrydium biforme*), and N.Z. silver pine (*Lagarostrobos colensoi* formerly *Dacrydium colensoi*). The last two species are the chief sources of the bicyclic diterpenoids manool and manoyl oxide (plus ketomanoyl oxide) respectively, and the chemical transformations of these compounds leading to ambergris-type odorants is discussed in detail in Chapter 11.

The resinols matairesinol (9) and α-conidendrin (10) comprise the principal constituents of the heart-shake resin of the mataī.[8] Both are classified chemically as lignans and each is related to podophyllotoxin (11), derivatives of which have applications in the clinical treatment of certain cancers.[81-84] Matairesinol itself has been shown to have anti-tumour action in cancer of mice. This compound also has antioxidant properties in fats, and possibly other related compounds from native plants should be examined for this use. In recent years workers at the University of Auckland[85] have succeeded in converting the dimethyl ether of matairesinol into a lignan related to steganacin (12), a compound which has pronounced therapeutic activity.

The heart-shakes of rimu are composed almost entirely of the diterpenoid podocarpic acid (13). Its structure is unusual in that it possesses only seventeen carbon atoms of the twenty normally found in diterpenoids. Many of its derivatives have pharmacological activity. The acid has considerable potential as a readily available source of an optically active starting material for transformation into commercially useful compounds, such as pharmacologically active steroids and perfume fixatives. Much research work in this area has been carried out overseas, particularly in ·the use of podocarpic acid as a starting point for the synthesis of new structures[86] and there has been considerable research in the same field by Cambie and co-workers at the

University of Auckland.[8] At Auckland, particular emphasis has been placed on the use of podocarpic acid for the preparation of intermediates for the production of optically active steroids or triterpenoids. However, the disposition of functional groups in the compound means that direct interconversions are difficult.

Since simple derivatives of podocarpic acid and other resin acids often show biological activity,[87-92] a logical approach to compounds with steroid-like activity has been to use the rigid diterpenoid skeleton as a supporting moiety for groups capable of interacting with physiological receptor sites. The introduction of a heterocyclic ring into a steroid molecule often produces new types of activity,[93-95] and thus the heterocyclic derivatives of podocarpic acid prepared by Cambie and co-workers[96] might be expected to show interesting biological activity. Use of the supporting skeleton approach has proved to be valuable in syntheses of compounds showing potential anti-neoplastic activity from a series of resin acids.[97]

1. Gardner, R. (1923) 6 *N.Z.J. Sci. Tech.* 147.
2. Gardner, R. (1924) 7 *N.Z.J. Sci. Tech.* 220.
3. Murray, J. (1949) 13 *J.N.Z. Inst. Chem.* 128.
4. Murray, J. (1950) 14 *J.N.Z. Inst. Chem.* 44.
5. Briggs, L.H. (1947) 80 *J. Roy. Soc. N.S.W.* 151.
6. Brooker, S.G., Cain, B.F., Cambie, R.C. (1963) 1 *Trans. Roy. Soc. N.Z.* 61.
7. Brooker, S.G., Cambie, R.C., James, M.A. (1966) 1 *Trans. Roy. Soc. N.Z.* 205.
8. Cambie, R.C. (1976) 6 *J. Roy. Soc. N.Z.* 307.
9. Cambie, R.C. (1988) 18 *J. Roy. Soc. N.Z.* 137.
10. Cain, B.F., Scannell, S., Cambie, R.C. (1961) 4 *N.Z.J. Sci.* 3.
11. Cambie, R.C., Cain, B.F., La Roche, S. (1961) 4 *N.Z.J. Sci.* 604, 707, 731.
12. Cain, B.F., La Roche, S., Cambie, R.C. (1962) 5 *N.Z.J. Sci.* 537.
13. Bailey, R.W., Pain, V. (1971) 10 *Phytochem.* 1065.
14. Russell, G.B., Fenemore, P.C. (1970) 13 *N.Z.J. Sci.* 61.
15. Calder, V.L., Cole, A.L.J., Walker, J.R.L. (1986) 16 *J. Roy. Soc. N.Z.* 169.
16. Brooker, S.G., Cambie, R.C., Cooper, R.C. *New Zealand Medicinal Plants* Auckland (Heinemann) 1987 268p.
17. Brooker, S.G., Cambie, R.C., Cooper, R.C. (1989) 43 *Econ. Bot.* 79.
18. Cambie, R.C. (1986) 2 *N.Z.J. Techn.* 111.
19. Buisson, D. (1979) Industrial Processing Division, DSIR, Information Series IPD I5/8.
20. Aston, B.C. (1917) 15 *N.Z.J. Agric.* 55, 117.
21. Aston, B.C. (1918) 16 *N.Z.J. Agric.* 358.
22. Aston, B.C. (1919) 17 *N.Z.J. Agric.* 136.
23. Aston, B.C. (1918) 1 *N.Z.J. Sci. Tech.* 264, 346.
24. Hummel, J.J. (1882) 1 *J. Soc. Chem. Ind.* (London) 170.
25. Nicholas, J.L. *Narrative of a Voyage in New Zealand in the years 1814 and 1815 in company with the Rev. Samuel Marsden*, 2 vols. 1817 431 395 p. Reprinted by Wilson and Horton, Auckland 1971.
26. Perkin, A.G. (1898) 73 *J. Chem. Soc.* 1019.

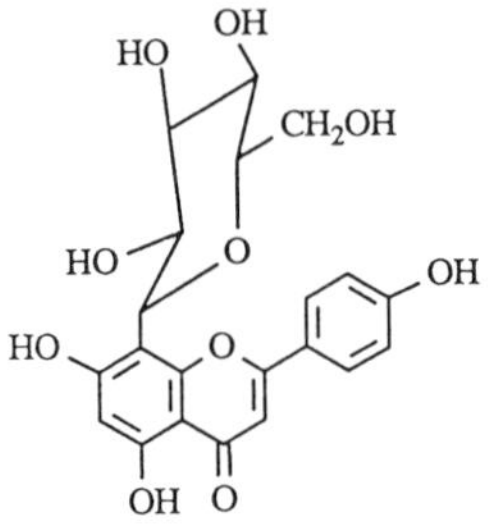

(1) vitexin

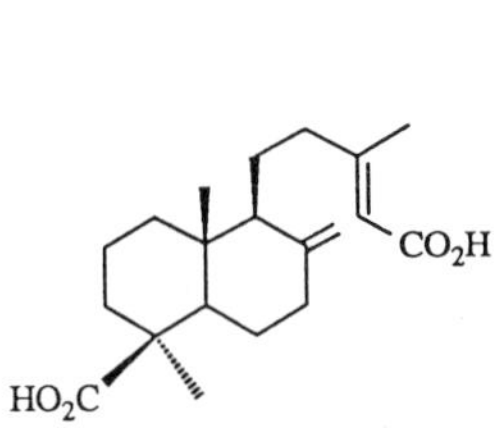

(2) agathic acid

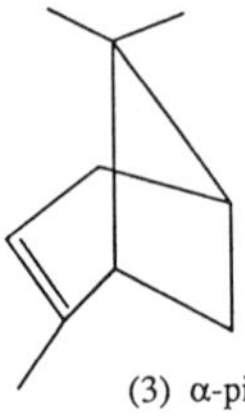

(3) α-pinene

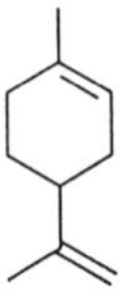

(4) limonene

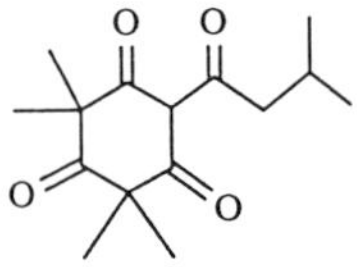

(5) leptospermone

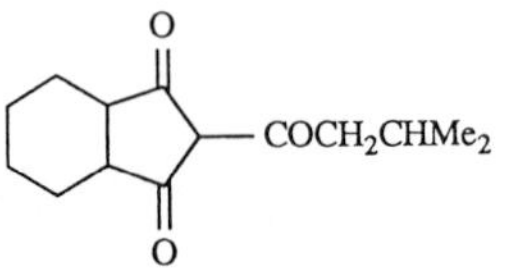

(6) valone

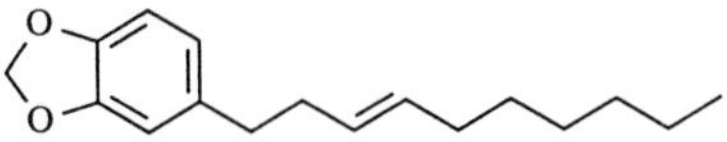

(7) juvadecone

27. Hughson, W.G.M., Ellis, A.J. *A History of the Chemistry Division* Wellington (N.Z. DSIR) 1981 201 p.
28. Wilson, R.D. (1979) 17 *N.Z.J. Bot.* 113.
29. Wilson, R.D. (1984) 22 *N.Z.J. Bot.* 195.
30. Thornton, G.G. *New Zealand's Industrial Heritage* Wellington (Reed) 1982 194p.
31. Kirk, T. *The Forest Flora of New Zealand* Wellington (Government Printer) 1889 345p, 142 pl.
32. Colenso, W. (1869, 2nd ed. 1875) 1 *Trans. N.Z. Inst.* 233.
33. Cranwell, L.M. *The Botany of Auckland. A Book for all Seasons* Auckland (Institute and War Memorial Museum Handbook) 1981 156p.
35. Hawley, G.G. (ed.) *Condensed Chemical Dictionary* 9th ed. New York (Van Nostrand Reinhold) 1977 957p.
36. Thomas, B.R. in *Organic Geochemistry*, New York (Springer Verlag) New York 1969 pp. 599, 618.
37. Hosking, J.R. (1929) 48 *Trav. Chim. Pays-Bas* 622.
38. Gough, L.J. (1964) *Chem. and Ind.* (London) 2059.
39. Thomas, B.R. (1966) 20 *Acta Chem. Scand.* 1074.

(8) a cyanogenic glyceride

R=C$_{17}$H$_{33}$ or C$_{17}$H$_{31}$

(9) matairesinol

(10) α-conidendrin

(11) podophyllotoxin

(12) steganacin

(13) podocarpic acid

40. Kauri Deposit Surveys Ltd (1979) Report on project concept and environmental assessment, lodged with Lands and Survey Department.
41. Anon. (1983) *Northland Age* 6 September, (1984) ibid. 10 April, 16 October, (1985) ibid. 18 June, 1 August, 5 November.
42. Irvine, J. (1983) *New Zealand Herald* 31 August.
43. Anon (1988) *New Zealand Herald* 9 February.
44. Cording, H.J. (1988) *Chem. and Ind. and Lab. Management* May, p. 24.
45. Scott, G. (1989) *New Zealand Herald* 7 April, p. 16.
46. Hingley, B. *Gumdiggers of the North. A Bulletin for Schools* Wellington (School Publications Branch, Dept. of Education) 1980 49p.
47. Reed, A.H. *The Gumdiggers. The Story of Kauri Gum* Wellington (Reed) 1972 193p.
48. Wagener, R. *Gumfields of Aupouri* Kaitaia (The Age) 1977 28p.

49. Briggs, L.H., Penfold, A.R., Short, W.F. (1938) *J. Chem. Soc.* 1193.
50. Briggs, L.H., Hassall, C.H., Short, W.F. (1945) *J. Chem. Soc.* 705.
51. Rawlings, G.B. (1958) 3 *N.Z. Soc. Soil Sci. Proc.* 41.
52. Hutchins, R.F.N., Sutherland, O.R.W., Gnanasunderam, C., Greenfield, W.J., Williams, E.M., Wright, H.J. (1984) 10 *J. Chem. Ecol.* 81.
53. Russell, G.B. and Fenemore, P.G. (1970) 13 *N.Z.J. Sci.* 61.
54. Russell, G.B. and Fenemore, P.G. (1971) 14 *N.Z.J. Sci.* 31.
55. Bowers, W.S., in Locke M., Smith D.S. (eds) *Insect Biology of the Future* New York (Academic Press) 1980 p. 621.
56. Nishida, R., Bowers, W.S., Evans, P.H., (1983) 1 *Arch. Insect Biochem. Physiol.* 17; Bowers, W.S., Nishida, R. (1980) 209 *Science* 1030.
57. Marker, R.E., Wagner, R.B., Ulshafer, P.R., Wittbecker, E.L., Goldsmith, D.P.J., Ruof, C.H. (1943) 65 *J. Am. Chem. Soc.* 1199.
58. Blunden, G., Jaffer, J.A., Jewers, K., Griffin, W.J. (1981) 44 *J. Nat. Prod.* 441.
59. Fankhauser, B.L., Brasch, D.J. (1985) 1 *N.Z.J. Tech.* 27.
60. McIlroy, R.J. (1949) *J. Chem. Soc.* 121.
61. McIlroy, R.J., Holmes, G.S., Mauger, R.P. (1945) *J. Chem. Soc.* 796.
62. Haslemore, R.M., Roughan, P.G. (1980) *DSIR Plant Physiology Division Report*, p.26.
63. Carter, C.L. (1943) 62 *J. Soc. Chem. Ind.* (London) 238 T; (1951) 2 *J. Sci. Food Agric.* 54.
64. Carter, C.L., McChesney, W.J. (1949) 164 *Nature* (London) 575.
65. Moyer, B.G., Pfeffer, P.E., Valentine, K.M., Gustine, D.L. (1979) 18 *Phytochem.* 111.
66. Miller, I.J., Furneaux, R.H. (1982) 25 *N.Z.J. Sci. Tech.* 15.
67. Briggs, L.H., Cambie, R.C. (1958) 2 *Tetrahedron* 256.
68. Briggs, L.H., Cambie, R.C., Candy, B.J., O'Donovan, G.M., Russell, R.H., Seelye, R.N. (1965) *J. Chem. Soc.* 2492.
69. Bick, I.R.C., Hai, M.A. (1981) 22 *Tetrahedron* 3275; (1981) 16 *Heterocycles* 1301.
70. Edgar, J.A., Eggers, N.J., Jones, A.J. and Russell, G.B. (1980) 21 *Tetrahedron Lett.* 2657.
71. Morice, I.M. (1962) 13 *J. Sci. Food Agric.* 666; (1967) 18 ibid. 129; (1969) 20 ibid. 611.
72. Brooker, S.G. (1960) 88 *Trans. Roy. Soc. N.Z.* 158.
73. Brooker, S.G. (1957) 84 *Trans. Roy. Soc. N.Z.* 935.
74. Morice, I.M. (1975) 14 *Phytochem.* 765.
75. Borges del Castillo, J., Brookes, C.J.W., Cambie, R.C., Eglinton, G., Hamilton, R.J., Pellitt, P. (1967) 6 *Phytochem.* 391.
76. Body, D.R. (1983) 60 *J. Am. Oil Chem. Soc.* 1894.
77. Morice, I.M. (1967) 18 *J. Sci. Food Agric.* 344; (1975) 14 *Phytochem.* 1315.
78. Palmer, J. (1985) 101 *Botany Division Newsletter* 20.
79. Princen, L.H. *Proc. Int. Conference on Oils, Fats, and Waxes, Fats for the Future* Auckland (Duromark Publishers) 1983 250p.
80. Brooker, S.G., Eyres, L. (1981) Paper delivered to N.Z. Inst. Chem. Conference, Auckland.
81. Hartwell, J.L., Schaer, M.J. (1947) 7 *Cancer Res.* 716.
82. Kupchan, S.M., Meshulan, H., Sneden, A.T. (1978) 17 *Phytochem.* 767.
83. Limburg, H. (1972) 22 *Planta Med.* 348.
84. Stahelin, H. (1972) 22 *Planta Med.* 336.

85. Cambie, R.C., Clark, G.R., Craw, P.A., Rutledge, P.S., Woodgate, P.D. (1984) 37 *Aust. J. Chem.* 1775.
86. Banerjee, A.K., Acevedo, J.C., Canudas-González, N. (1990) 99 *Bull. Soc. Chim. Belg.* 9.
87. Baizer, M.M., Karnowsky, M., Bywater, W.G. (1950) 72 *J. Am. Chem. Soc.* 3800.
88. Bible, R.H. (1956) United States Patents 2750270, 2750373, 2753357, 2759014; (1957) 2786431, 2796431; (1958) 2851485, 2854474; (1960) 2971008.
89. Bible, R.H., Hoehn, N.W.M. (1956) United States Patent 2705725; (1959) 2889371.
90. Picha, G.N. (1956) United States Patent 2767162.
91. Brandt, C.W., Ross, D.J. (1948) 161 *Nature* (London) 892.
92. Fieser, L.F., Campbell, W.P. (1939) 61 *J. Am. Chem. Soc.* 2528.
93. Martin-Smith, M., Khatoon, T. (1963) 6 *Prog. Drug Res.* 279.
94. Matsumoto, T., Ohsuga, Y., Harada, S., Fukui, K. (1977) 50 *Bull Chem. Soc. Japan* 266.
95. Mori, K., Matsui, M. (1970) 26 *Tetrahedron* 3467.
96. Beresford, G.D., Cambie, R.C., Mathai, K.P. (1973) 26 *Aust. J. Chem.* 1763.
97. Rodina, L.B., Bril, G.E., Dubkova, N.I. (1970) *Sin. Prod. Kanifoli Skipidara* 281 (*Chem. Abs.* 1972, 76, 141080).

11

Fragrances and flavours

11.1 Essential Oils

Essential oils are obtained by distilling plant material, usually the leaves, with water; the oil which separates from the distillate often has a characteristic odour identified with the individual species or genus. In the days of the alchemists, such a fragrant oil was thought to be the essence of the plant — hence the name 'essential oil'. The term 'oils' is also applied to petroleum products, not relevant to this book and to non-volatile oils which occur generally in the seeds of plants. The latter are sometimes referred to by the antiquated term 'fixed' oils but since they are mainly esters of glycerol and fatty acids, glyceride oils is a better term (see Chapter 10).

Essential oils have been known from the Middle Ages, and for countless years have made up the principal constituents of flavours and fragrances. The most abundant components of essential oils are terpenes, such as geraniol (see figure formula 1) from the oil of geranium, citronellal (2), which is responsible for the characteristic aroma of lemon oil, menthol (3) in peppermint oil, and camphor (4) from the camphor tree. Also present in essential oils are sesquiterpenes, C_{15} compounds, such as farnesol (5), which has the characteristic odour of lily of the valley, and bisabolene (6), which is found in the oils of bergamot and myrrh.

In addition to providing raw materials for the perfumery and flavouring industries, most essential oils have bactericidal or disinfectant properties. In Australia, 'Tea-tree oil' from *Melaleuca alternifolia* has strong germicidal properties and was, for many years, produced commercially by the Australian Essential Oil Co. Ltd in Sydney under the name 'Tri-trol'. It was reputed to be useful for cuts, tinea, mouth ulcers, and so on, and also in the treatment of gonorrhoea and diabetic gangrene.[1] An eventual decline in the production of tea-tree oil was presumably due to an unreliable supply, inconsistent quality, and a lack of promotion. However, even now, it is still being produced and sold in pharmacies and health food shops throughout Australia.

The isolation of essential oils is one of the easiest of extraction procedures. In its simplest form, the plant material is placed in a vessel fitted with a lid and an outlet pipe which, in turn, is connected to a water-cooled condenser. After the addition of water, the leaves are boiled and the steam, enriched with vapour of the oil, is passed through the condenser, where it is re-liquified. The oil, which is immiscible and lighter than water, moves to the top of a receiver, where it may be collected. A refinement is to pass steam, generated from a separate boiler, through the charge of plant material.

The analysis of essential oils from New Zealand plants has been studied for many years, especially at the University of Auckland and the University of Otago. Initial studies were carried out using simple fractionating columns which required large amounts of the oil, and which usually led, after repeated fractional distillation, to the identification of only the major constituents. However, since about 1945, with the construction of highly efficient fractionating equipment, detailed analyses of many essential oils from native plants have been carried out.[2] With the advent of gas liquid chromatography, many minor components can be detected, and this has resulted in a re-examination of some essential oils investigated in earlier years. The coupling of gas chromatographs to mass spectrometers has made the analysis of essential oils a much easier task than in former times. However, despite these technical advances, the chemical analysis of essential oils has become unfashionable in New Zealand laboratories. This is possibly because the commercial exploitation of essential oils in a country where there is little chemical tradition and where labour costs are high, make any programme a risky economic business. Nevertheless, an American businessman has expressed interest in the possibility of New Zealand becoming a major supplier of essential oils[3] and a joint venture programme is about to commence with a horticultural firm on the East Cape.

The most comprehensive study of essential oils in New Zealand has been carried out on the native gymnosperms, of which there are twenty species distributed among ten genera. Of those which have been investigated, all have yielded between 0.1 to 2% of oil and most oils contain large amounts of terpenes ($C_{10}H_{16}$), small amounts of terpene alcohols, aldehydes, and esters, and *c.* 40% of sesquiterpenes ($C_{15}H_{24}$).[4] The terpene fractions invariably contain α-pinene (7), which is the most widely distributed of all terpenes and, in addition, often contain limonene and myrcene. The most characteristic feature, however, is the occurrence of diterpenes ($C_{20}H_{32}$) throughout, often in high yield, and frequently with each species affording a different isomer as the major component. The kaikawaka (*Libocedrus bidwillii*) is unique among the New Zealand pines in yielding an oil which contains no diterpenes but instead, contains a mixture of solid paraffins.[5,6]

Few other groups of New Zealand plants are noted for the production of

mānuka (*Leptospermum scoparium*) **Myrtaceae**
Painting: E. Blumhardt

essential oils. The largest group of importance is the Myrtaceae which, except for the mānuka (*Leptospermum scoparium*), has not been examined in detail. In Australia, the family includes important oil-bearing trees such as *Eucalyptus* and *Melaleuca*. Mānuka is probably the most widespread tree or shrub of the native flora, and its essential oil is probably the most valuable, although it has not been used commercially. Its oil is pleasant smelling, and could well be used as an inexpensive perfume for soaps. The constituents are highly oxygenated, and besides the terpene hydrocarbons α-pinene, eudesmene, and aromadendrene, are comprised of citronellal, citronellol, citral, geraniol, cineole, and an unusual ketonic compound leptospermone (8)[2] (see Chapter 10). The related species kānuka (*Kunzea ericoides*) is not as common as mānuka, but it produces a similar type of oil.

A survey of the oils from the rātā (*Metrosideros* species) has been made by Gardner[7] and a more detailed examination would be of interest both chemically and botanically. The species fall into two groups on the basis of their essential oil constituents, one group yielding oils containing terpenes and sesquiterpenes, together with their oxygenated derivatives, and the other giving oils which contain mainly sesquiterpenes and sesquiterpene alcohols.

Among other essential oils which have been investigated in some detail[2] are those of *Pittosporum eugenioides* and *P. tenuifolium*, horopito

(*Pseudowintera colorata*), *Olearia paniculata* and *O. ilicifolia*, and kawakawa (*Macropiper excelsum*). As indicated in Chapter 9, the essential oil of the last plant contains the aromatic ether myristicin (9), which accounts for the use of this plant as a 'painkiller' for toothache.

As the late Professor L.H. Briggs pointed out in 1946 in a Liversidge Research Lecture to the Royal Society of NSW,[8] comparatively little has been done in the synthetic field using terpene constituents as raw materials, despite the fact that α-pinene is one of the cheapest of naturally occurring chemicals. He suggested there was considerable scope for industrial use of terpenes. Most had the initial advantage of being optically active and, by the introduction of hetero atoms, oxygen, nitrogen, sulphur, and so on, a wide diversity of products could be obtained and exploited. Today terpenes are in fact exploited to advantage in many overseas industrialized countries. However, while many individuals have expressed interest in the utilization of New Zealand natural products, especially essential oils, no concerted effort has ever been made to develop these products.

While it is not a native species, brief mention should be made of *Pinus radiata* as a source of tall oil and turpentine. In recent years Pine Chemicals Ltd has installed tall oil and turpentine plants at Mt Maunganui.[8] Tall oil, a by-product from the kraft pulping mills of Tasman and Forest Products, is injected into heated columns and the feed stock split into vapors and residue. The process is repeated many times until the desired degree of refining has been achieved. Tall oil provides fatty acids, rosin, and pitch, the refined products of which can be used in the manufacture of alkyd resins, soaps, detergents, emulsifiers, surfactants, resins, adhesives, paper size, and printing inks. A batch process as opposed to a continuous process is employed to fractionate the turpentine mainly into α- and β-pinenes, which can be used for the manufacture of flavourings, fragrances, disinfectants, solvents, mining chemicals, and rubber processing aids.

11.2 Fragrances

Included in this section are those essential-oil bearing plants which might have value in the perfumery field. Chief among these are members of the family Rutaceae which, in contrast to Australia, is represented in this country by only a few species. Mairehau (*Phebalium nudum*) is a once common but now rare aromatic shrub of northern New Zealand (Plate 16), and is related to the popular hedge plant *P. squameum* (formerly *P. billardieri*) of Australia. The leaves of mairehau were used by the Māori as perfume. The sweet scent of this plant is due to an essential oil which was shown by C.B. Radcliffe and W.F. Short[10] to contain terpinyl acetate, limonene, camphene, citronellal, citral, sesquiterpenes, and a sesquiterpene alcohol. *Melicope ternata* (whārangi) and

M. simplex, the other two members of the Rutaceae in New Zealand, are also becoming rare. They contain small amounts of essential oils. The former is one of the most pleasant smelling trees of the flora, but the oil was obtained by Radcliffe[11] on too small a scale for detailed investigation. It is currently under re-investigation at the University of Auckland.

In earlier times a number of other plants were used by the Māori for their perfumery properties. Colenso[12] reported that

> A gum resin, used [by Māori] to perfume their oil [for anointing], was obtained from the kohuhu and tarata (*Pittosporum tenuifolium* and *P. eugenioides*), and also from the taramea (*Aciphylla colensoi*) which was very highly prized. The strong smelling ferns, *Hymenophyllum villosum*, *Doodia media*, and *Phymatodes* [now *Phymatosorus*] *scandens* were also used for the same purpose of perfuming and for scenting oil; and so were a few fragrant mosses and liverworts, especially *Lophocolea novae-zelandiae* and *L. allodonta*. The aromatic leaves of raukawa (*Pseudopanax edgerleyi*) were also sought for a similar purpose, particularly to rub on limbs and bodies. The daisy-like flowers of the roniu (*Brachycome radicata*), and the flowering tops of the sweet-scented grass kāretu (*Hierochloe redolens*), were worn around the neck, enclosed in fibrous leaves, as a scented necklace.

Martin[13] reported that

> The hepaticae [liverworts] are not known to have any commercial uses; but the old-time Māori women used to gather a few of the larger species of *Lophocolea* and *Chiloscyphus* because of the fragrance which they imparted to their anointing oils. Small sachets of the same liverworts were often worn around their necks.

The scent of the root parasite, *Dactylanthus taylorii*, a native flowering plant in the tropical family, Balanophoraceae, has been described as 'delicious daphne-like', 'like a ripe melon' and 'decidedly pleasant when the flowers first expand, but becomes heavy and disagreeable when they commence to decay'. The parasite is common on the roots of many native trees on the Volcanic Plateau and is boiled to obtain the 'wooden rose', the fluted wood of the host root. The smell, when boiling is under way, is indescribably bad.

Another notorious scent is that of the ball or net fungus, sometimes called the basket fungus (*Clathrus cibarius*) found most years in damp mānuka scrub, waste places, and old gardens. The Māori names of these two plants are significant; *Dactylanthus* was known as pua reinga (flower of Hades) or waewae-atua (fingers or toes of the spirit), while *Clathrus* was known as tūtae kēhua (ghost dung). This latter bad smell is said to attract flies which spread the fungal spores. Back in the early 1930s someone left a parcel of the fungus for the Auckland War Memorial Museum botanist, Lucy Cranwell, and an attendant stowed it away in a safe corner. Before Lucy Cranwell returned several weeks later, the Botany Hall was uninhabitable, and the drains were undergoing rigorous examination.

In recent years, an Auckland firm has marketed perfumes under the names 'kowhai', 'rata', and so on. It is not known if any indigenous oil is included in these formulations; the producers may have been using the names as a novel marketing ploy. However, positive efforts are being made in Opunake, Taranaki, to incorporate perfume essences from the petals of native plants such as kōwhai, rewarewa, and clematis in the production of toilet soaps.[14]

'Jonathan Spade'[15] has recently brought to the attention of readers of his gardening column in the *New Zealand Herald* the virtues of fragrant native species. In addition to some of those mentioned above he also drew attention to mānuka (*Leptospermum scoparium*), New Zealand honeysuckle or rewarewa (*Knightia excelsa*), mangeao (*Litsea calicaris*), māhoe (*Melicytus ramiflorus*), kaikomako (*Pennantia corymbosa*), *Olearia odorata* and its close relatives, scented sundew (*Drosera binata*), the native *Celmisia* species, and *Cordyline australis* and its relatives. All of these plants have perfumed flowers and 'Spade' made the plea for a New Zealand enthusiast to list appropriate descriptions and cultivation notes for them.

One fragrant plant of possible commercial value is the scented fern *Paesia scaberula*, the finely lacing fronds of which exude a sweet perfume, especially in warm sunlight. Its essential oil was investigated by L.H. Briggs and M.D. Sutherland[16] who showed the presence of a sesquiterpene alcohol and the paraffin, n-heptacosane. Scientists at Forest Research Institute, Rotorua have given consideration to using *P. scaberula* as an understorey crop in wide-spaced *Pinus radiata* stands on sites which are unsuitable for agroforestry. The ferns would have the role of controlling weeds without competing with the pines, and at the same time providing a source of compounds with prospects in perfumery. The fern, which is a weed pest in pastoral farming, grows under other species (mānuka, for example), is beneficial in soil conservation, excludes other species, and grows to only about 40-50 cm high.[17]

11.3 Perfumes Based on Ambergris Substitutes

One of the most valued of the animal perfumes is ambergris, a metabolic product of the sperm whale, *Physeter macrocephalus*, which accumulates as a concretion in the gut of the animal. It can be obtained directly from the intestines of the whale but is occasionally found washed ashore as compact brown or grey pieces ranging in weight from several hundred grams to several kilograms. Some controversy still exists as to how ambergris is formed in the whale;[18] current ideas are that it is only formed in old or diseased whales and then only when the animals are feeding on squid or cuttlefish. It has been suggested that ambergris is produced as a result of irritation of the whale's stomach by the indigestible beaks of the squid and cuttle-fish.[19] Ambergris has been prized for hundreds of years for its unique fragrance and fixative power, and in Asia it has been claimed to have medicinal and aphrodisiac properties. In the form of a tincture, it homogenizes and exalts most perfume oils, giving them, as a result of its fixative capacity, a tenacity which surpasses most other tinctures in the musk class. Because of its restorative qualities it enjoyed great popularity in medieval Europe and it played a prominent role in the development of high-quality French perfumes. Unfortunately, in recent times excessive hunting of the sperm whale has resulted in the virtual disappearance of ambergris from the world market* and in order to meet the needs of the perfumery industry, chemists have sought to provide synthetic or semi-synthetic equivalents.

The composition of ambergris has been known for many years[19-23] and it has been found that the odorous components amount to less than 0.03% of the crude material.[24] The characteristic odour of ambergris results from the presence of appreciable amounts (up to 74% of the extract)[25] of the tricyclic triterpene alcohol ambrein (10). Ambrein itself is odourless and the long-lasting odour property of ambergris probably results from a constant autoxidation of ambrein at the central double bond of the molecule with the formation of odorous products.[19] A study by Ohloff[26] has shown that many of the odorous compounds found in ambergris[27] can be produced in the laboratory by oxidation of ambrein. This has led to the suggestion that singlet oxygen from the air is an active reagent in the biodegradation of ambrein, especially since porphyrins, which are known to be efficient photosensitizers, have also been identified in ambergris.

The elucidation of the structure of ambrein by Swiss chemists in 1947,[28] and its subsequent degradation into odorous compounds showed that, in general, ambergris-type odorants have the *trans*-decalin skeleton which is present in many plant diterpenoids.[29] As a result of the establishment of a relationship between ambreinolide (11), one of the major oxidation products of ambergris, and the plant diterpenoids manool (13) and sclareol (14), considerable interest in manool has been aroused both in New Zealand and abroad. Although there are several sources of the compound, New Zealand holds a unique position in that pink pine (*Halocarpus biformis*) has the highest known content (up to 6-8%) based on the dry material weight. Extensive work has been carried out on the production of synthetic substitutes for odoriferous compounds such as

---

* It is reputed that two Northland fishermen recently found a large lump of ambergris on the shore near the Kaipara Harbour. Realizing its importance they had it sent to France by the New Zealand agents of a French perfumery company. They were somewhat surprised but overjoyed to receive a cheque for $187,000.

ambreinolide starting from manool or its related compounds manoyl oxide (15) and 2-oxomanoyl oxide (16). The latter two compounds occur as the major diterpenoids in New Zealand silver pine (*Lagarostrobos colensoi*). Much of this work has been carried out within New Zealand by groups led by Professors R.C. Cambie and P.K. Grant, at the University of Auckland and the University of Otago respectively.

One of the most significant developments was the degradation of manool by potassium permanganate or osmium tetroxide-sodium periodate oxidations to the epimeric acetals (17) and (21), the former of which is reputed to have the strongest ambergris-type odour yet known. *These acetals have been used in the preparation of a wide variety of perfumes although there is no speciality product made from either substance. A semi-industrial production of these acetals from manool, reported by Demole[31] in 1964, heightened interest in the latter compound and a number of commercial firms within New Zealand have attempted to export manool or its relatives to perfumery companies in Europe and North America.

Initially, in the early 1960s, a price of $16 per pound was offered for crude manool, and an Auckland firm supplied some 500lbs as a trial shipment, this material being successfully employed in perfume manufacture.[32] However, the firm lost its advantage when its European customer Firmenich et Cie balked at an increase in price. A survey by the New Zealand Forest Service at about this time showed that *Halocarpus biformis* grew mainly on the West Coast of the South Island and that its total volume was at least 750,000 cubic feet, representing 10,200 tons of dry timber from which 800 tons of manool could be produced. At price of $10 per pound (in 1964) for the compound, the gross value would have been about $6 million.[33] Fortunately, the supply has never had to be used on this basis.

Methods for large scale extraction of manool from pink pine were investigated by members of the Chemistry Division of DSIR Petone, in the early 1960s.[34] Early attempts were not entirely successful, but later, two alternative methods were developed.[33] One involved solvent extraction followed by vacuum co-distillation of the extract with glycerol and centrifugal separation of crystalline manool from the mother liquors. The second method involved radio frequency capacitive heating of the wood chips in a temperature-controlled chamber under vacuum to distil the manool directly from the chips. Little came of the second method, but the firm of U.E.B.

---

* Until recently the internal acetal (21) was believed to have a strong ambergris-type odour. However, Ohloff and co-workers[30] have shown that the acetal is perceived by most people as practically odourless and they attribute the earlier erroneous odour-evaluation to contamination of samples by traces of the powerfully odorous epimer (17).

Industries Ltd in Auckland took up the commercial production of manool using the first extraction technique. Despite initial success and some patented research on the development of the internal acetals from manool,[35] a decreasing market and increased costs saw U.E.B. cease production after only a few years of operation.

Meanwhile, work within the New Zealand universities and in overseas laboratories on analogues of the odoriferous acetals was continuing, manool becoming one of the starting materials for the investigation of structure–activity relationships in this particular series of odorants. Most of the current theories concerned with the phenomenon of odour postulate a relationship between type of odour and molecular structure. Bicyclic compounds possessing the odour of ambergris usually have a *trans*-fused decalin skeleton and many possess the tricyclic oxide type structure which is found in the naturally occurring diterpenoid manoyl oxide (15). Variation in the molecular geometry near the oxide group can lead to a pronounced effect on the odoriferous properties, and much research has been carried out to define more closely the structure–activity relationships. In the ensuing discussion, only the work carried out within New Zealand is highlighted.

Other acetals of importance which have been derived from manool are the intramolecular ortho esters (18) and (22)[36] and the noracetals (19) and (23).[37] Cambie and Palmer[38] established that as for the other isomeric pairs, only the C 13 α-substituted epimer (19) possessed a strong ambergris-type odour. The nitrogen derivatives (20) and (24) have been prepared recently and their odours are 'being further evaluated',[39] Grant having shown earlier that the oxathiolan (25) possessed a very weak woody odour which was inferior to that of the oxygen analogue (17).[40] Compounds in the ent-15, 16-dinor series, (26) and (27), have odour strengths far below those of their epimers (17) and (21), and in addition, the pair (26) and (27) show little difference in odour quality other than a strong anosmia effect.[41] A loss of intensity with unchanged tonality occurs on introducing a double bond in the molecule as in the compounds (28)[42,43] and (29).[44] This is also the case when going from the dehydrogenated compound (30) to the saturated ether (12).[45,46] Inversion of the acetal group as in (31) led to no change in the odour quality[47] while substitution of the C 13 methyl group by a hydrogen atom results in the loss of the odour properties of the nor compound.[48]

A number of other compounds derived from manool have also been assessed for their odour characteristics.[49,50] For example, manool can be converted in a single stage reaction into the lactone (32) which has an odour similar to that of ambergris.[51] Full details of this have been recorded in a review by Vlad.[52] Interest in the derivatives of manool was heightened with the discovery that compounds such as manoyl oxide were present in certain strains of Turkish tobacco.

Although seventeen compounds have been identified in the neutral fraction of a heartwood extract of the silver pine (*Lagarostrobos colensoi*), the principal constituents, as indicated earlier, are manoyl oxide (15) and 2-oxomanoyl oxide (16). Cambie and co-workers have degraded the latter compound to make labdanyl ethers substituted in the 2-position by oxygen. Such compounds belong to the same odour type as their unsubstituted analogues[46] but have weaker and somewhat sweeter odours. In contrast, derivatives which are oxidized at C 6 are odourless.[44] Manoyl oxide can be converted into the unsaturated lactone (33) which is used by the Reynolds Tobacco Co. in the US as a tobacco-flavouring agent[53] and it has also been converted into the odoriferous five-membered oxide (34)[54] known commercially as Ambrox®. The enol ether (36) is an important intermediate in the preparation of perfumery fixatives.[55] The latter compound has been synthesized from manoyl oxide via oxidative decarboxylation of the acid (38), and the corresponding keto-oxide (37) has been obtained in a similar manner from 2-oxomanoyl oxide.[46] These and other oxidation products have served as important intermediates in the synthesis of perfumes.

The methods of synthesis of the oxides (34) and (35) (Ambroxide®) from sclareol (14) were the first in this area to be patented and numerous subsequent modifications of individual stages have been made in order to adapt them to industrial conditions. Recently, Cambie and Palmer[38] have prepared a series of five-, six-, seven-, and eight-membered ring dioxa compounds which are analogues of the commercially available odorants Homofixateur (12), Ambrox® (34) and Isoambrox (39). All were shown to possess odours of the ambergris type, the odour decreasing in intensity as the size of the dioxa ring was increased. This work has been extended to include demethyl analogues of Ambrox® and Isoambrox, starting from podocarpic acid (40)[56] which is readily available from rimu (*Dacrydium cupressinum*) (see Chapter 10).

During the last ten to fifteen years there has been a considerable growth of interest in the perfumes obtained from labdanoids such as manool, manoyl oxide, and 2-oxomanoyl oxide, as evidenced by the increasing number of publications on the subject. The interest is due in part to the fact that these compounds are present in tobacco, imparting to it special organoleptic properties, and to the fact that addition of these compounds to various kinds of tobacco improves its quality. Their value for perfumery has made the synthesis of their degradation products not only of academic but also of practical interest. Not only do they have highly effective properties as perfumes; they also possess a complex odour which embraces a wide range of different shades. The discovery that oxygenated analogues of Ambrox® with comparable odours can be prepared from the readily available podocarpic acid opens up further potential for the development of a fine chemical industry in New Zealand. One of the present authors expressed the opinion in 1971 'that it will be necessary to produce an odoriferous derivative on a semi-industrial scale before our raw materials such as manool, manoyl oxide, and ketomanoyl oxide are going to be of much value to New Zealand'.[32] As yet, no commercial attempts have been made in this direction, and a chance by an enterprising firm to establish a perfumery industry in New Zealand goes begging.

11.4 Flavourings

Colenso[12] reported that the Māori people were often curiously particular as to what plants were used, tied around, or under and over their vegetable food, in their cooking ovens in the earth. For instance, the roots of the tī koraha (*Cordyline pumilio*)* were tied separately for baking in bundles of hangehange † (Geniostoma ligustrifolium); for their kai, or prepared sweet potatoes, they used the leaves of parataniwha (*Elatostema rugosum*); generally, however, they used the fronds of the larger ferns Blechnum *'capense'*‡ and *Pneumatopteris pennigera*. The Te Kao Māori people are still particular as to what goes in the earth oven and where, and an elder always supervises an oven. The Māori women prepare the food, but the elder and his helpers make the fire, prepare the oven, and do the loading so that mixed flavours are controlled, not random.

The Rev. Canon Walsh[57] in 1901 described the Māori use of the prepared roots of tī, or tī pore (*Cordyline terminalis*) and tī rauriki* (*C. pumilio*) and discussed their flavours. Tī pore was sweeter, and was used as a sweetmeat. A.H. Watt of Te Kao[58] thought it might have been used in an oven to give a sweet flavour to 'pork' (human or porcine). H. Beattie,[59] an authority on the southern Māori people, wrote in 1920 that, in the old days, the root of the tī or cabbage tree, known as 'kauru', was cooked in an umu or oven, or at a fire, and eaten by the Māori people. It tasted its best when soaked in flax-honey (wai-korāri). If you were travelling and thirsty, you could upturn your kauru root and let the moisture trickle down your throat — it was a good sweet drink.

The use of the sweet juice of tūpākihi berries (tutu, *Coriaria*) to sweeten seaweed extract and prepared bracken root appears to be another case of using a flavour (cf Chapter 8); seaweed on its own is not very interesting, although Cranwell[60] describes the use of seaweeds, collectively known as 'rimu', as relishes in cooking at Bethells Beach, especially in earth ovens used for large gatherings. All of the early writers to whom reference is made mention the

---

* Both Tīkoraha and Tīrauriki are Māori names for *Cordyline pumilio*.
† Hangehange (*Geniostoma ligustrifolium*) often gives northern bush a characteristic smell.
‡ Kiokio (*Blechnum capense*) lacks a specific name at present as capense is illegitimate.

sweetness of tutu berry juice (while stressing the need for care in its preparation owing to its poisonous origin). The use of tītoki (*Alectyron excelsus*) for the preparation of a genuinely local product, a tītoki liqueur, is almost the sole commercial attempt to utilize the flavouring potential of New Zealand plants. However, as mentioned in Chapter 8, distinctive New Zealand honeys, produced primarily from the flowers of native species, are marketed mainly for overseas consumption.

1. Lassak, E.V., McCarthy, T. *Australian Medicinal Plants* Australia (Methuen) 1983 240p.
2. Cambie, R.C. (1976) 6 *J. Roy. Soc. N.Z.* 307.
3. Ualesi, F. (1989) *New Zealand Herald* 25 August, p. 58.
4. Murray, J. (1949)13 *J. N.Z. Inst. Chem.* 128; (1950) 14 ibid. 44.
5. Batt, R.D., Hassall, C.H. (1949) 68 *J. Soc. Chem. Ind.* (London) 359.
6. Batt, R.D. Slater, S.N. (1949) *J. Chem. Soc.* 858.
7. Gardner, R. (1931) 50 *J. Soc. Chem. Ind.* (London) 141T.
8. Briggs, L.H. (1947) 80 *J. Roy. Soc. N.S.W.* 151.
9. Hanger, D. (1983) 17 *Chem. Ind. Lab. Man.* Oct/Nov 7.
10. Radcliffe, C.B., Short, W.F. (1928) 47 *J. Soc. Chem. Ind.* 324 T.
11. Radcliffe, C.B. (1929) 60 *Trans. N.Z. Inst.* 251.
12. Colenso, W. (1869, 2nd ed. 1875), 1 *Trans. N.Z. Inst.* 233.
13. Martin, W. *The Flora of New Zealand* Christchurch (Whitcombe and Tombs) 1961 (4th ed.) 335p.
14. Macdonald, I. (1988) *New Zealand Herald* 28 December, p. 62.
15. Spade, J. (1985) *New Zealand Herald* 28 September.
16. Briggs L.H., Sutherland, M.D. (1947) 160 *Nature* (London) 333.
17. Franich, R.A. Pers. commun. 1985.
18. Dubois, P.A. (1978) *Parfums cosmet. Aromes* 35 (*Chem. Abs.* 1978, 88, 197414).
19. Lederer, E. (1950) 6 *Fortschr. Chem. Org. Naturst.* 87.
20. Belov, V.N., Skvortsova, N.I. (1963) 32 *Russ. Chem. Rev.* 121.
21. Collin-Asselineau, C., Lederer, E., Mercier, D., Polonsky, J. (1950) *Bull. Soc. Chim. Fr.* 720.
22. Lederer, E. (1949) *J. Chem. Soc.* 2115.
23. Ohloff, G. (1969) 12 *Fortschr. Chem. Forsch.,* 185.
24. Ruzicka, L., Seidel, C.F., Pfeiffer, M. (1948) 31 *Helv. Chim. Acta* 827.
25. Korzh, L.N., Strigina, L.I. (1972) *Maslob. Zhir. Prom.* 23 (*Chem. Abs.* 1973, 78, 33795).
26. Ohloff, G., Schulte-Elte, K.H., Muller, B.L. (1977) 60 *Helv. Chim. Acta* 2763.
27. Jegou, E., Polonsky, J., Lederer, E., Schulte-Elte, K.H., Egger, B., Ohloff, G. (1977) 1 *Nouveau J. Chim.* 529.
28. Jeger, O., Durst, O., Buchi, G. (1947) 30 *Helv. Chim. Acta* 1853.
29. Cambie, R.C. (1967) 3 *Chem. Ind. N.Z.* 4.
30. Ohloff, G., Vial, C., Wolf, H.R., Job, K., Jegou, E., Polonsky, J., Lederer, E. (1980) 63 *Helv. Chim. Acta* 1932.
31. Demole, E. (1964) 20 *Experientia* 609.
32. Cambie, R.C. (1971) 35 *Chem. in N.Z.* 69.
33. Merz, D.F., Ritchie, W.J. (1970) 13 *N.Z.J. Sci.* 268.
34. McDonald, I.R.C. (1964) 1 *Chem. Ind. N.Z.* 16.

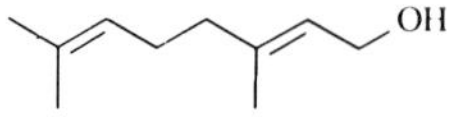

(1) geraniol

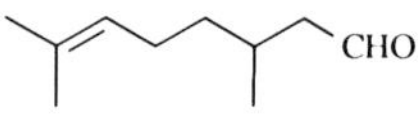

(2) citronellal

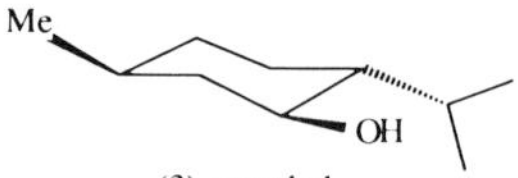

(3) menthol

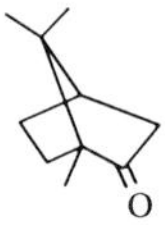

(4) camphor

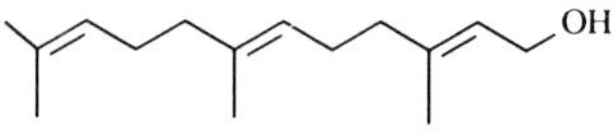

(5) farnesol

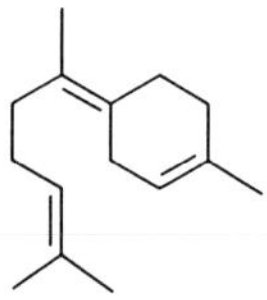

(6) bisabolene

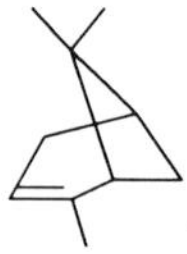

(7) α-pinene

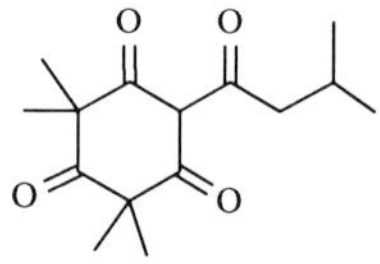

(8) leptospermone

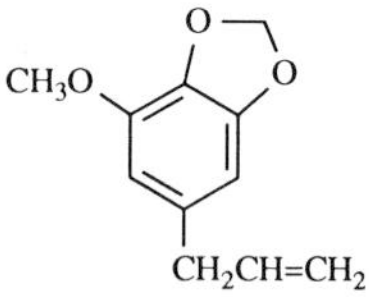

(9) myristicin

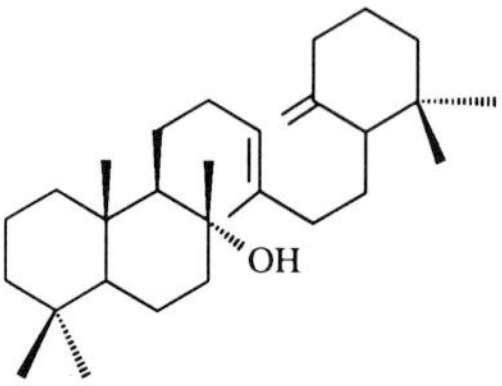

(10) ambrein

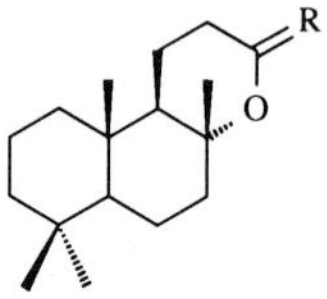

(11) R = O ambreinolide
(12) R = H$_2$ homofixateur

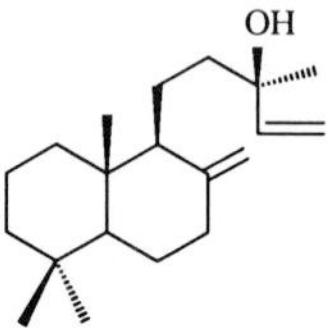

(13) manool

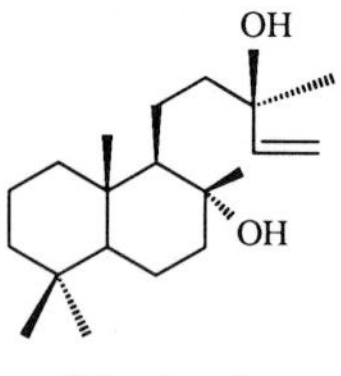

(14) sclareol

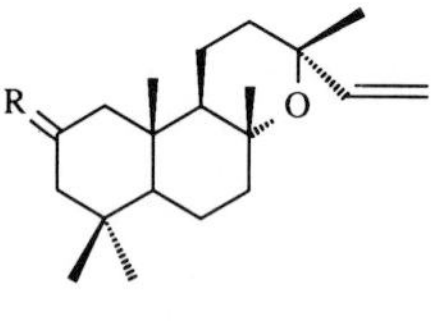

(15) R = H$_2$ manoyl oxide
(16) R = O 2-oxomanoyl oxide

35. Nicholls, M.J. (1973) U.E.B. Industries Ltd. N.Z. Patent 167959.
36. Demole, E., Wuest, H. (1967) 50 *Helv. Chim. Acta* 1314.
37. Grant, P.K., Liau, H.T.L., Temple, W.A. (1979) 32 *Aust. J. Chem.* 1353.
38. Cambie, R.C., Palmer, B.D. (1981) 34 *Aust. J. Chem.* 1265.
39. Grant, P.K., Prasad, J.S., Rowan, D.D. (1983) 36 *Aust. J. Chem.* 1197.
40. Grant, P.K., Liau, H.T.L. (1978) 31 *Aust. J. Chem.* 1777.

(17) R = Me
(18) R = OMe
(19) R = H
(20) R = NHMe

(21) R = H
(22) R = OMe
(23) R = H
(24) R = NHMe

(25)

(26)

(27)

(28)

(29)

(30)

41. Dey, A.K., Wolf, H.R. (1978) 61 *Helv. Chim. Acta* 1004.
42. Grant, P.K., Liau, H.T.L. (1978) 31 *Aust. J. Chem.* 1791.
43. Grant, P.K., Liau, H.T.L., Nicholls, M.J. (1973) 26 *Aust. J. Chem.* 1815.
44. Sandermann, W., Bruns, K. (1966) 99 *Chem. Ber.* 2835.
45. Bruns, K., Sandermann, W. (1968) 24 *Tetrahedron* 1357.
46. Cambie, R.C., Joblin, K.N., McCallum, N.K. (1970) 23 *Aust. J. Chem.* 1439.
47. Cambie, R.C., Joblin, K.N., Preston, A.F. (1972) 25 *Aust. J. Chem.* 1767.
48. Grant, P.K., Rowan, D.D. (1979) 32 *Aust. J. Chem.* 1395.
49. Grant, P.K., Lai, C.K. (1978) 31 *Aust. J. Chem.* 1785.

(31) R = H$_2$
(32) R = O

(33)

(34) R = H$_2$ Ambrox®
(35) R = O Ambroxide®

(36) R = H$_2$
(37) R = O

(38)

(39) Isoambrox®

(40) podocarpic acid

50. Ohloff, G., in Theimer, E.T. (ed.) *Fragrance Chemistry, the Science and the Sense of Smell* New York (Academic Press) 1982 635p.
51. Cambie, R.C., Joblin, K.N., Preston, A.F. (1971) 24 *Aust. J. Chem.* 2365.
52. Vlad, P.F., (1982) 51 *Russ. Chem. Rev.* 644.
53. R.J. Reynolds Tobacco Co. (1963) Fr. Patent 1346170 (*Chem. Abs.* 1964, 60, 14475).
54. Cambie, R.C., Joblin, K.N., Preston, A.F. (1971) 24 *Aust. J. Chem.* 583.
55. Ohloff, G., Giersch, W., Schulte-Elte, K.H., Vial, C. (1976) 59 *Helv. Chim. Acta* 1140.
56. Cambie, R.C., Palmer, B.D. (1982) 35 *Aust. J. Chem.* 601.
57. Walsh, Rev. Canon (1901) 33 *Trans. N.Z. Inst.* 301.
58. Watt, A.H. Pers. commun. (1975).
59. Beattie, H. (1920) 52 *Trans. N.Z. Inst.* 66.
60. Cranwell, L.M. *The Botany of Auckland. A Book for All Seasons* Auckland (Auckland Institute and War Memorial Museum Handbook) 1981 156p.

12

Solanum Alkaloids

New Zealand possesses three indigenous *Solanum* species, poroporo or kohoho (*S. aviculare*), cut-leaved nightshade (*S. laciniatum*), and small-flowered nightshade (*S. nodiflorum*). According to Baylis,[1] *S. aviculare* is also found in the Kermadec Islands, on Lord Howe Island, and in Queensland, while *S. laciniatum* is also located in Tasmania. Each plant grows in Victoria and the southern part of New South Wales and South Australia.

In recent years these species, and especially *S. aviculare*, have assumed increasing economic significance as sources of suitable chemical materials for the industrial production of hormonal steroids, such as progesterone (the pregnancy hormone), oestrone (the female hormone), and testosterone (the male hormone). Work carried out within New Zealand on the structural aspects of the steroidal glycoalkaloids which can be isolated from the plants, and the attempted commercial utilization of these compounds in this country, warrants a full chapter of this book. Indeed, the utilization of the New Zealand species, particularly in Eastern Europe, illustrates the potential of certain native plants, in this case a potential which was largely unrecognized and certainly unexploited in New Zealand until overseas interests forced the facts upon us. The importance accorded to *S. laciniatum* in Russia is reflected in the issue of a postage stamp in 1972 depicting this plant (Plate 10).

The two major groups of steroids manufactured on an industrial scale in the US and Europe are contraceptive hormones and the anti-inflammatory agents derived from corticosteroids. As a direct result of the dramatic rise in the sales of anti-fertility agents in recent years, increasing demands have been placed on the supply of the naturally occurring steroids from which these drugs are prepared. The total synthesis of steroids is a complex and expensive process, albeit an ever-increasing area of industrial research, and therefore the continued supply of steroidal drugs relies heavily on the availability of naturally occurring compounds. At present, the major raw material for the manufacture of steroidal drugs is diosgenin, a sapogenin which occurs as a glycoside in Mexican yam (*Dioscorea mexicana*), a plant indigenous to Central America. The aglycone can be converted into 3β-acetoxypregna-5, 16-dien-20-one in 65-68% overall yield, and this latter product is then used for the manufacture of a wide range of steroid drugs. Nationalization of the supply of diosgenin by the Mexican government in 1974 and the subsequent rapid increase in its cost, together with an increasing world-wide demand for steroidal raw materials, has provoked a search for alternative starting materials.[2] One such material is the alkaloid solasodine, which is the nitrogen analogue of diosgenin, and which can be obtained by hydrolysis of the glycosidic alkaloids isolated from many *Solanum* species. The species *S. aviculare* and *S. laciniatum* (Plates 10 and 11) are particularly important in this respect since they produce the alkaloid in high yield.[3] According to Carl Djerassi, a noted American chemist, 'solasodine may well become the diosgenin in the next decade although it is unlikely that any major steroid manufacturers will want to become dependent again on any single source'.[2] Up until 1981 steroidal alkaloids have been isolated from over 350 *Solanum* species; these include seventy-five steroidal alkamines whose structures are all known.[4]

Chemical interest in *Solanum* alkaloids began about 165 years ago, when Monsieur Desfosses, a pharmacist at Besançon in eastern France carried out an examination of black nightshade (*Solanum nigrum*) with the aim of isolating the narcotic principle.[5,6] Despite the well-known poisonous nature of the plant, especially to stock which ate it, the plant had long been used in medicine as a narcotic, sudorific, anti-inflammatory agent, and as a gargle and mouthwash. It was also used in the treatment of burns, ulcers, ringworm, gout, and earache.[7] Desfosses isolated an impure alkaloid which he named 'solanine'. This is now known to be mainly solamargine.[8] Desfosses also examined other *Solanum* species, isolating from bittersweet (*S. dulcamara*) a glycoalkaloid which he recognized as different from 'solanine' and which he called 'dulcarine'.[6]

A number of investigations of various *Solanum* species followed, including that in which solanine was isolated from the common potato (*S. tuberosum*)[9] and that of an alkaloid which was later to be recognized as solasonine from apple of Sodom (*S. sodomeum*, now known as *S. linnaeanum*).[10-13] At this time, however, the latter compound was called 'solanine-s' to distinguish it from 'solanine-t', derived from *S. tuberosum*. Despite considerable work by the Italian chemist Oddo during the early 1900s the structure of solasonine remained obscure. A further glycoalkaloid was isolated in 1939 from the fruits of the Indian medicinal plant *S. xanthocarpum* by Saiyed and Kanga.[14] They called this compound 'solancarpine', not recognizing it to be identical with 'solanine-s' (solasonine).

In the view of Louis and Mary Fieser, the authors of a classic book on

Professor L.H. Briggs, D.Sc., D.Phil., F.R.S.N.Z.

steroids,[15] 'modern work in the [Solanum] field was initiated in 1942 by Briggs at Auckland, New Zealand'. The worker referred to was the late Professor Lindsay Heathcote Briggs (1905-1975), affectionately known to his colleagues as 'Bob' Briggs, who was then carrying out research work on the chemistry of the native flora which was to earn him international recognition; a recognition hitherto gained by only a handful of New Zealand chemists. Working with few resources and without the benefit of modern equipment, Briggs isolated 'solanine-s' from *S. auriculatum* now known as *S. mauritianum* (woolly nightshade),[16] showed it was identical with 'solancarpine',[17] and recommended that, henceforth, the glycoalkaloid be known as solasonine and its aglycone as solasodine.[18] In 1939, Briggs[19] and Rochelmeyer,[20] an American chemist, independently proposed the correct molecular formula $(C_{27}H_{43}NO_2)$ for solasodine, and Briggs suggested that solasodine was a hydroxy derivative of solanidine, the aglycone of the potato aklaloid.[21]

In 1942, Bell and Briggs[22] isolated solasonine from the New Zealand species *Solanum aviculare* and suggested two possible structures for the aglycone. However, it was not until 1950, some 130 years after the work of Desfosses, that Briggs and his co-workers[23] finally proposed the correct structure for solasodine. This was confirmed three years later by the American chemist, Uhle, who achieved a partial synthesis of the compound from the steroidal sapogenin kryptogenin.[24] Sato and co-workers[25] first reported the degradation of solasodine into 16-dehydropregnenolene acetate in 1951, and a number of methods for this conversion have been published subsequently, there having been considerable competition between various groups of workers to improve both the procedures and yields for the various steps.[26-30] The best modern procedure is detailed in Scheme 1. In a continuous operation from solasonine without isolation and purification of the intermediates, an overall yield of 65-68% of the final product, 3β-acetoxypregna-5, 16-dien-20-one, can be obtained. Another method[31] which is based on earlier work by Briggs[32] is given in Scheme 2, while an improved sequence from O-acetyl-N-nitrososolasodine[33] is given in Scheme 3.

The initiative and early work on the use of *Solanum* species as a source of raw material for steroidal production began in Hungary, where the collection of medicinal plants has long been a traditional industry. By 1962, the commercial production of solasodine from *S. aviculare* had been commenced in the Akaloida Chemical Factory at Fiszavasvari.[34] *Solanum laciniatum* had been cultivated in botanical gardens in Europe as a decorative plant since the first visit to New Zealand by Captain Cook in 1769-70, at which time Mr (later Sir) Joseph Banks and Dr Daniel Solander had collected specimens to take back to London. *Solanum aviculare* was found by the Forsters on Cook's second voyage, and described as an esculent plant by G. Forster in 1786 (see Chapter 2). The early cultivation of these plants in Hungary was from single berries growing in a European botanical garden,[35] but this was later augmented by the collection of fresh material from New Zealand in the early 1960s. In the ensuing years, much work has been carried out on the development and cultivation of stable, high-yielding strains of *S. aviculare* and *S. laciniatum*. Most of this work has been carried out in Hungary and the USSR. Such studies include investigations on the genetics and breeding, induced mutation, seed germination, general growing conditions, and resistance of the plants to pests and diseases. In addition, reports have been published on the tissue culture of the plants, their hydroponic culture, and on the harvesting and storage of the plant material.[34] Recent reports suggest that solasodine is assuming increasing importance as a source of pregnane derivatives in the commercial synthesis of steroids.[36,37]

Many chemists have addressed themselves to the study of the extraction of solasonine from *S. aviculare* and *S. laciniatum*, and thence the hydrolysis to solasodine. On a laboratory scale, the extraction is relatively simple, but some problems occur on scaling up the operation. Chief among these is the darkening of the product obtained from dried plant material and the need to minimize the formation of unwanted solasodiene during acid hydrolysis. These problems have been solved satisfactorily in Hungary and the USSR, where production has been carried out for some time. The difficult handling of the glycoalkaloid can be avoided by direct hydrolysis of the finely ground plant material with aqueous isopropanol containing 5% of hydrochloric acid,[38,39] and a five-tank cascade reactor system has been designed for continuous operation using this procedure.[40] Besides Hungary and the USSR, where the production of solasodine is of the order of 20-30 kg per hectare per year and

5 kg per hectare per year respectively, industrial production of solasodine from
S. aviculare or *S. laciniatum* is actively being carried out in Romania and the
People's Republic of China. In addition, growth trials on these species and/or
extraction of the glycoalkaloids on a pilot plant scale have been carried out in
Pakistan, Egypt, Israel, Japan, England, Indonesia,[34] and possibly other
countries.[41]

In New Zealand, cropping trials were begun in the early 1960s by Ivon
Watkins-Dow Ltd at New Plymouth,[42] and were later taken up by the Crop
Research Division of the DSIR in Canterbury. *Solanum aviculare* yields up to
2.5% of solasodine from the dried aerial parts of the plant, and it has been
reported[43] that *S. laciniatum* contains approximately half as much, although
in both cases the ultimate yield of solasodine is dependent on the climate and
soil conditions. The berries and young shoots of these plants contain a
somewhat higher proportion of the alkaloid. A detailed investigation of
methods for the isolation and hydrolysis of the glycoalkaloids has been carried
out in the Chemistry Division of the DSIR at Petone.[43] However, none of the
reported methods appear to offer any great improvement over those
introduced more than twenty years ago by Briggs and his co-workers.[44,45] In
1976, a company, Solanum Extraction Industries Ltd, was set up in Waitara,
Taranaki, to grow and process *S. aviculare* and *S. laciniatum* to produce
solasodine. The company was jointly owned by Fletcher Holdings Limited
and Diosynth Netherlands BV, a subsidiary of Akzopharma, the
pharmaceutical division of the Dutch firm AKZO NV. The later firm is a bulk
supplier of alkaloids, biochemicals, and steroidal hormones.[46]

A full plant was commissioned in 1978 and growing was concurrently
expanded from that required for pilot plant production. Unfortunately,
although high hopes were held for the successful commercial exploitation of
the crop, the company ceased production in 1981. Reasons suggested for this
decision have been the difficulty of reproducing a satisfactory and economic
yield of the alkaloid from the plant and an undercutting of the market by
material produced in the People's Republic of China. Whatever the precise
reason, the failure of this venture does not augur well for similar New Zealand
industries based on natural products. Despite the obvious advantages of an
indigenous plant source and a history of local chemical endeavour in the field,
it might appear in hindsight that New Zealand was far too slow to capitalize on
what might have been a very lucrative industry. Of interest is the fact that a cost
estimate of the production of solasodine from *S. aviculare* carried out in
Australia[34] at about the same time as the New Zealand commercial venture
was starting did not lead to commercial production.

Apart from their use as raw materials for the steroidal hormone industry,
the glycosidic steroidal alkaloids may be useful in their own right. All of the
glycosidic steroidal alkaloids tested so far, and some of their alkamines, show

antibiotic activity against certain fungi and bacteria.[41] For example, reports from Poland record the control of hair diseases in guinea pigs and rabbits,[47] ringworm in chinchillas,[48] and skin disease of cattle.[49] Such reports would accord with the early use by the Māori of all parts of the plants for scabies, and as a healing salve. In clinical trials, the aglycone solasodine has been found to effect improvement in stomach and liver functions with minimal side effects[50] and to show cortisone-like activity in humans.[51-54] Solasodine has also been shown to possess cardiac and anti-inflammatory effects in mice and to show some effects on the vascular permeability as well as on the central nervous system.[41] Its anti-accelerator cardiac action and that of its glycosides was first shown by Krayer and Briggs.[55] The repellent action of some *Solanum* glycoalkaloids on Colorado beetle (*Leptintoarsa decemlineata*) and its larva has been studied in detail,[56-62] and a relationship between repellency and molecular structure partly established. Although they are inactive against the larva of houseflies,[63] two Asian potato beetles,[45] and tobacco hornworm,[64] they are active against potato leafhopper (*Epoasca fabae*).[65] Of interest in this respect is the recent work of Cambie and co-workers at the University of Auckland, who have successfully converted solasodine into a nitrogen-analogue of an ecdysone.[66] Ecdysones are naturally-occurring compounds with a steroidal nucleus, which have a promising future as agents for the control of insects. The background to these studies has been discussed in detail by Cambie and Potter,[45] and only a brief account is recorded here.

Insects, crustaceans, and other arthropods have tough cuticles or exoskeletons which serve to support and contain their internal organs and muscles. As these cuticles are incapable of growth or modification, they are periodically discarded and replaced by new ones. This process of moulting or 'ecdysis' allows the animal to grow and acquire adult characteristics. In many insects, such as moths, butterflies, and beetles, these moults can be accompanied by extensive structural modification as the animal passes through the chrysalis or pupa stage into the winged insect. Such remarkable transformations have fascinated man from time immemorial and have inspired and enriched folklore. In 1922 it was suggested that a hormonal mechanism was responsible for initiating moulting in insects, and in 1954 the herculean task of isolating and characterizing the moulting hormone (named α-ecdysone) from the silkworm (*Bombyx mori*) was accomplished.

Exposure of an insect to ecdysones can induce it to moult at inappropriate times, thereby disturbing its normal development and resulting in death. Ecdysones thus have a promising future for insect control for several reasons. Firstly, they are active in extremely low concentration. Secondly, their action is not limited to specific stages and periods of insect development, but they are able to produce destructive aberrations in the moulting process at any time during metamorphosis, and also in the adult stage. Thirdly, they act non-specifically on insects and arthropods, but they are virtually inactive in other animals; and in vertebrates, including man, they are apparently harmless.

Although ecdysone and several related compounds have been obtained from a variety of insects and crustaceans, much greater quantities can be isolated from plants, particularly the bark and leaves of New Zealand conifers of the family Podocarpaceae (see Chapter 10). New Zealand ferns, particularly members of the Blechnaceae, also contain phytoecdysteroids. To date, over forty naturally occurring moulting hormones (phytoecdysones) have been isolated from plant sources, and all have been found to be steroids which are related structually to the ecdysone molecule. Undoubtedly they are part of the plant's defensive mechanism. By examining the effect of structure of the ecdysones on their moulting hormone activity, it has been possible to assess the relative importance of various structural elements of the molecule for biological activity. The most important property of many phytoecdysones is their enormous hormonal activity compared with that of α-ecdysone. Some are up to twenty times more active. At such concentrations they can have a very damaging effect on insect development, partly because they resist the inactivation undergone when α-ecdysone is applied to insects. For example, while α-ecdysone is 50% inactivated within seven hours of being fed to insects, cycasterone, a phytoecdysone from *Cyas* is only inactivated to the same extent after thirty-two hours.[69] Also, it has been shown that certain synthetic steroids with only minimal structural similarity to the ecdysones have been found to inhibit moulting and metamorphosis severely in the development of insects, that is they can act as ecdysone antagonists. Following the observation by Robbins and his co-workers[67,68] that steroids with a nitrogen atom in the sidechain (aza steroids) inhibited insect growth, Cambie and co-workers have successfully converted solasodine into an ecdysone-analogue in which the nitrogen atom of the alkaloid is retained in the product. Such a compound gives rise to the possibility of a new type of hormonal insecticide which, on application, will inhibit growth and development in all larval stages. One may expect that, as well as the high activity of such compounds, an additional advantage will be the inability of insect pests to adapt to these insecticides. Again, while overseas firms have expressed interest in the New Zealand research it has gone virtually unnoticed in this country.

1. Baylis, G.T.S. (1954) 82 *Trans. Roy. Soc. N.Z.* 639.
2. Djerassi, C. (1976) 195 *Proc. Roy. Soc. London* B 175.
3. Briggs, L.H., Cambie, R.C. (1958) *J. Chem. Soc.* 1422.
4. Ripperger, H., Schreiber, K., In Rodrigo, R. (ed.) *The Alkaloids Vol. XIX* New York (Academic Press) 1981 227p.
5. Desfosses, (1820) 6 *J. Pharm.* 374.
6. Desfosses, (1821) 7 *J. Pharm.* 414.
7. Greive, M. *A Modern Herbal* 2 vols, New York (Hafner Publishing Co.) 1971.

Scheme 1

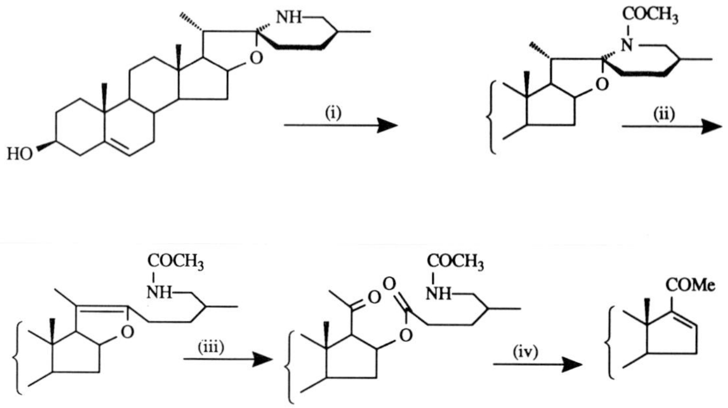

(i) Ac₂O/Py   (ii) HOAc   (iii) CrO₃/HOAc   (iv) HOAc/Δ

Scheme 2

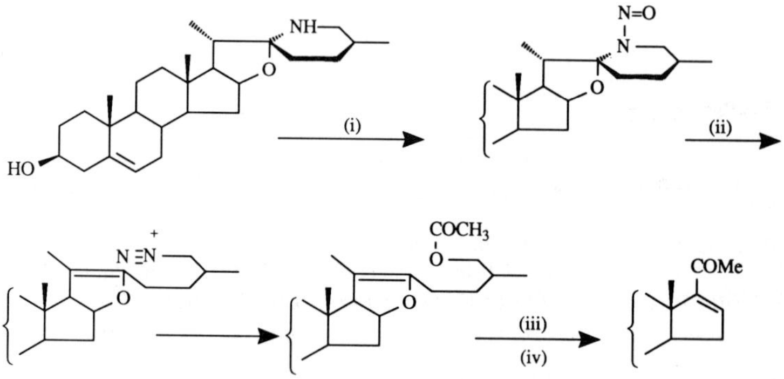

(i) HNO₂   (ii) HOAc/NaOAc/Ac₂O   (iii) CrO₃/HOAc   (iv) HOAc/Δ

8. Prelog, V., Jeger, O. In Manske, R.H.F. and Holmes, H.L. (eds) *The Alkaloids Vol. III* New York (Academic Press) 1953 p. 248.
9. Baup, M. (1826) 31 *Ann. Chim.* 108.
10. Missaghi, G. (1875) 5 *Gazz. Chim. Ital.* 417.
11. Oddo, G., Colombano, A. (1905) 35 *Gazz. Chim. Ital.* 27.
12. Romeo, G. (1905) 35 *Gazz. Chim. Ital.* 579.
13. Soldanini, A. (1905) 44 *Boll. Chim. Farm.* 769, 808, 843.
14. Saiyed, I.Z., Kanga, D.D. (1936) 44 *Proc. Ind. Acad. Sci.* 255.
15. Fieser, L.F., Fieser, M. *Steroids* New York (Reinhold) 1959 945p.

Scheme 3

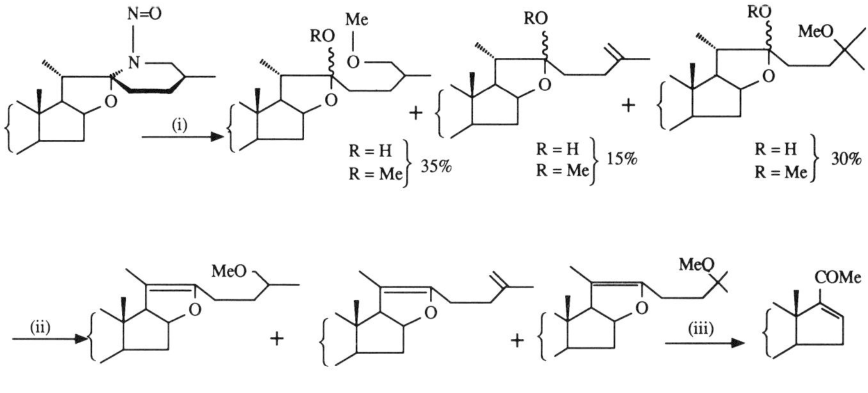

(i) TsOH/MeOH   (ii) Ac₂O/Δ   (iii) Na₂Cr₂O₇

16. Healy, A.J. (1974) *Proc. 27th N.Z. Weed Pest Control Conf.* 41.
17. Briggs, L.H. (1937) 59 *J. Am. Chem. Soc.* 1404.
18. Briggs, L.H. (1937) 59 *J. Am. Chem. Soc.* 2467.
19. Briggs, L.H. (1939) 144 *Nature* (London) 247.
20. Rochelmeyer, H. (1939) 277 *Arch. Pharm.* 329.
21. Clemo, G.R., Morgan, W. McG., Roper, R. (1936) *J. Chem. Soc.* 1299.
22. Bell, R.C., Briggs, L.H. (1942) *J. Chem. Soc.* 1.
23. Briggs, L.H., Harvey, W.E., Locker, R.H., McGillivray, W.A., Seelye, R.N. (1950) *J. Chem. Soc.* 3013.
24. Uhle, F.C. (1953) 75 *J. Am. Chem. Soc.* 2280.
25. Sato, Y., Miller, H.K., Mosettig, E. (1951) 73 *J. Am. Chem. Soc.* 5009.
26. Bite, P., Tuzson, P. (1958) 17 *Acta Chim. Acad. Sci. Hung.* 241.
27. Labler, L., Cerny, V. (1961) *Tagungsber. Dtsch. Akad. Landwirtschaftswiss* (Berlin) 207.
28. Magyar, G. (1961) *Tagungsber. Dtsch. Akad. Landwirtschaftswiss* (Berlin) 225.
29. Sato, Y., Ikekawa, Y., Mosettig, E. (1959) 24 *J. Org. Chem.* 893; (1960) 25 ibid. 783.
30. Suvorov, N.N., Sokolova, L.V., Morozovskaya, L.M., Mursasheva, V.S. (1958) 3 *Khim. Nauka i Prom.* 281 (*Chem. Abs.* 1958, 52, 20248).
31. Schreiber, K., Ronsch, H. (1963) *Tetrahedron Lett.* 937.
32. Sato, Y., Latham, H.G., Briggs, L.H., Seelye, R.N. (1958) 79 *J. Am. Chem. Soc.* 6089.
33. Bakker, C.G., Vrijhof, P. (1978) *Tetrahedron Lett.* 4699.
34. Collins, D.J. (1977) *Proc. 3rd Asian Symp. Medicinal Plants and Spices*, Colombo, Sri Lanka, UNESCO Publ. 42.
35. Tetenyi, P. (1965) 14 *Acta Agron. Acad. Sci. Hung.* 245.
36. Franz, C., Jatisatieur, A. (1983) 123 *Dtsch. Apoth. Ztg.* 1669 (Chem. Abs. 1983, 99, 67469).
37. Sree, A., Rao, Y.R., Mahapatra, S.N. (1982) 27 *Res. Ind.* 236 (*Chem. Abs.* 1983, 98, 215891).

38. Panina, V.V., Gorodetskii, L.Sh., Pimenova, T.B., Pogrebnaya, L.F., Babanova, A.V. (1972) 6 *Khim.NFarm. Zh.* 31.
39. Panina, V.V., Gorodetskii, L.Sh., Demyanova, G.M., Babanova, A.V. (1975) 9 *Khim.-Farm. Zh.* 34.
40. Rozanov, L.S., Lakona, M.I., Kondrat'ev, S.A. (1974) 8 *Khim.-Farm. Zh.* 26.
41. Schreiber, K., In Manske, R.H.F. (ed.) *The Alkaloids, Chemistry and Physiology Vol. X* New York (Academic Press) 1968 668p.
42. Cambie, R.C., (1971) 35 *Chem. in N.Z.* 69.
43. Weston, R.J. (1976) 26 *J. Appl. Chem. Biotech.* 657.
44. Briggs, L.H., Brooker, E.G. (1958) *J. Chem. Soc.* 1419.
45. Cambie, R.C., Potter, G.J. (1979) 43 *Chem. in N.Z.* 91.
46. Kloosterman, J. (1979) 43 *Chem. in N.Z.* 146.
47. Kamyszek, F. (1974) 11 *Zwierzeta Lab.* 57 (*Biol. Abs.* 1976, *61*, 8739).
48. Kamyszek, F. (1971) 17 *Herba Pol.* 195.
49. Chwojnowski, A., Kamyszek, F., Peisert, W. (1972) 18 *Herba Pol.* 34.
50. Pevchikh, V.V., Trusov, V.V., Belasludtsev, I.A. (1966) 38 *Ter. Arkh.* 48.
51. Daroczy, A., Hernadi, F. (1971) 6 *Acta Biochem. Biophys. Acad. Sci. Hung.* 327.
52. Leschinskii, L.A., Trusov, V.V., Pevchikh, V.V., Belasludtsev, I.A. (1968) 5 *Vrach Delo* 47.
53. Leschinskii, L.A. Pevchikh, V.V. Trusov, V.V., Belasludtsev, I.A. (1968) 31 *Tr. Izhevsk. Med. Inst.* 37 (*Biol. Abs.* 1969, 50, 103398).
54. Pevchikh, V.V. (1967) 54 *Ref. Zh. Otd. Yyh. Farmakol. Khimioter Svedstra Toksikol.* 410 (*Biol. Abs.* 1967, 48, 111589).
55. Krayer, O., Briggs, L.H. (1950) 5 *Br. J. Pharmacol.* 118, 517.
56. Buhr, H. (1961) *Tagungsber. Dtsch. Akad. Landwirtschaftswiss* (Berlin) 309.
57. Buhr, H., Toball, R., Schrieber, K. (1958) 1 *Entomol. Exp. Appl.* 209.
58. Fraenkel, G.S. (1959) 129 *Science* 146; (1961) *Tagungsber. Dtsch. Akad. Landwirtschaftswiss* (Berlin) 297.
59. Kuhn, R., Low, I. (1957) 69 *Angew.* Chem. 236; (1961) *Tagungsber. Dtsch. Akad. Landwirtschaftswiss* (Berlin) 7; (1961) 94 *Chem. Ber.* 1088.
60. Schreiber, K. (1957) 27 *Zuechter* 289; (1958) 1 *Entomol. Exp. Appl.* 28.
61. Sturchow, B. (1959) 78 *Biol. Zentrallb.* 142.
62. Sturchow, B., Low, I., (1961) 4 *Entomol. Exp. Appl.* 133.
63. Bergmann, E.D., Levinson, Z.N., Mechoulam, R. (1958) 2 *J. Insect Physiol.* 162.
64. Yamamoto, R.T., Fraenkel, G. (1960) 53 *Ann. Entomol. Soc. Am.* 499.
65. Dahlman, D.L., Hibbs, E.T. (1967) 60 *Ann. Entomol. Soc. Am.* 732.
66. Cambie, R.C., Potter, G.J., Read, R.W., Rutledge, P.S., Woodgate, P.D. (1981) 34 *Aust. J. Chem.* 599.
67. Svoboda, J.A., Robbins, W.E. (1971) 6 *Lipids* 113.
68. Thompson, M.J., Serban, N.N., Robbins, W.E., Svoboda, J.A., Shortino, T.J., Dutky, S.R., Cohen, C.F. (1975) 10 *Lipids* 615.
69. Harborne, J.B. *Introduction to Ecological Biochemistry* London (Academic Press) 1988 (3rd ed.) 131p.

13

Marine Macroalgae*

13.1 Rimurapa

Rimurapa (*Durvillaea antarctica*), also known as bull kelp, is a brown alga which grows between tidemarks on the most exposed rocks, to which it is attached by a disc-shaped leathery holdfast. The blade is honeycombed with air spaces that make it buoyant, and the Māori people made bags (pōhā) to store food and oil, by slitting the air layer between the two outer leathery surfaces.

A pōhā tītī was a bag made of kelp or tōtara bark, in which mutton birds (tītī) were preserved in their own fat. Sometimes a flax kit (kete) was plaited around the kelp bag of mutton birds for carrying or storage. Bags made of the fronds of rimurapa are still used in the mutton bird industry and to carry seaweeds.[1,2]

13.2 Fodder and Fertilizers

Although Dieffenbach (1843)[3] noted that pigs on the seashore ate certain kinds of seaweed, little has been done to investigate their suitability as animal feed. For some years, a meal prepared from an imported brown alga, *Ascophyllum nodosum*, has been available to New Zealand farmers as an animal additive; in Tasmania however, a relative of rimurapa, *Durvillaea potatorum*, is harvested and part of the crop is manufactured into animal nutritional supplements.

During the First World War, when supplies of potash could not be imported from Germany, B.C. Aston (1916) drew attention to the work of Dr E.J. Russell, Director, Rothamsted Experimental Station, England, who recommended that farmers utilize sources of potash other than the Stassfurt salts, such as ashes of seaweed.

---

* The use of seaweeds as food is discussed in Chapter 8 (Food and Beverages).

Locally, a suggestion was made that New Zealand's returned soldiers be provided with light work collecting and burning seaweed for potash, but Aston considered that more information was needed regarding available supplies of seaweed, its identity, and chemical composition. 'Returned soldiers' in 1916 were men who had been invalided back to New Zealand because of wounds or illness, and Aston doubted whether the work was 'light' or suitable for them. He concluded that the most economical way to utilize seaweed as a fertilizer was to give it a preliminary drying on the beach, and then to plough it into land near the sea; its heavy weight made transport inland too expensive. He thought that it would be suitable for potatoes, garden vegetables, and mangels, and that it was with potash-loving crops such as these, and with reclaimed swamp soils deficient in potash, that further experiments might be worthwhile.[4]

There was another flurry of research during the Second World War, when the Botany Division of the DSIR was asked to make a general survey of seaweeds. Lucy Moore reported in 1941 on the economic value of seaweeds generally, and on local species which might provide substitute products. Reference should be made to that report, reprinted in 1966, for detailed information on the considerable potential of seaweeds.[5]

Before the outbreak of war in 1939, New Zealand imported about 16,000 tons of potash fertilizers annually, the main sources of supply being France and Germany. With the stopping of these supplies, and the loss of much shipping, an inter-departmental committee was set up to investigate possible local sources of supply. Representatives met from the Department of Agriculture, Geological Survey, Botany Division, and Dominion Laboratory, and kelp (*Macrocystis pyrifera*) was chosen for study. Kelp was collected at Marlborough Sounds, dried, finely ground, and found to contain 17% of potash and 2% nitrogen. Higher yields of soluble potassium salts were obtained in subsequent chemical analyses. The growth of kelp plants was studied, and the kelp beds of Cook Strait and Foveaux Strait, the only beds in New Zealand likely to be economic, were mapped and measured.

From cutting and drying trials and chemical analyses, it was estimated that it would cost £9 10s a ton to produce seventy-five tons of dried kelp per week in a factory operated by ten men. It was also estimated that the Cook Strait and Foveaux Strait beds could provide about 5000 tons of dried kelp each year, yielding perhaps 1200 tons of potassium chloride. The possible use of fish wastes as a supplement to the ground kelp was also costed.[6]

Top-dressing trials at Waihi, and in North Taranaki and Southland, where it was known that potash could be profitably used on grassland, indicated that 17% of potash in ground kelp produced almost as much grass growth as 40% potash from imported salts. Meantime, in the market place, importers had obtained sufficient potash from Palestine (now Israel) and sulfate from America to meet the needs of New Zealand farms throughout the war years.[7]

In 1946, V.J. Chapman began various algal projects at the Botany Department, University of Auckland. Co-workers included E.C.S. Little (1948), who studied the rates at which buried seaweeds lose chemicals, and R.I.B. Francki (1960, 1964), who examined the manurial values of a brown alga (*Durvillaea antarctica*), and a red alga (*Pachymenia himantophora*, now *P. lusitoria*).[8-10]

Francki's analyses showed that *Durvillaea* is richer than *Pachymenia* in cations and chloride, but poorer in nitrogen and sulphur. When applied to soil as a dried meal, *Pachymenia* stimulated the growth of tomato seedlings on some soils and depressed it on others, while *Durvillaea* depressed growth on all soil types used. Both seaweeds, and particularly *Durvillaea*, were found to immobilize inorganic nitrogen in soils. Subsequently, *Pachymenia* meal was found to greatly increase the availability of manganese, which became toxic to plant growth on acid soils. The results suggest that the manurial value of a seaweed cannot be judged solely on its mineral analysis.

After the war, seaweed fertilizers with the trade names 'Maxicrop' and 'Seagro' became available; both are manufactured in New Zealand from partially treated *Ascophyllum nodosum*, imported from Britain. 'Maxicrop' is primarily used for gardens and glasshouse crops, while 'Seagro' and similar liquid manures are mainly promoted for pastures, orchards, and field crops.

In mid 1980, Growth Marketing Ltd of Auckland launched 'Response', a liquid seaweed fertilizer made locally from an extract of *Pterocladia* produced at the Opotiki factory of Coast Biologicals Ltd, with the nitrogen, potassium, and phosphorus elements boosted to levels of 9%, 4%, and 6% respectively.

Today, seaweed fertilizers seem to fall into four categories: 1) composts for gardens; 2) extracts for 'irrigating' pastures, market garden crops and so on; 3) extracts containing growth-promoting substances for the treatment of seeds before planting, and 4) extracts of a similar nature for spraying or wiping on the leaves of plants, especially indoor pot-plants. Claims regarding growth-promoting substances such as the cytokinins have been the subject of controversy, and the manufacturers of 'Maxicrop' took legal action for damages following criticism of their product.

13.3 *Pterocladia* and Agar

At the outbreak of war in 1939, when supplies of agar from Japan were threatened, the Botany Division of the DSIR began a study of substitutes. Supplies of two red algae, *Pterocladia lucida* and *P. capillacea*, were found to be adequate, and a trial batch of agar, prepared at the Dominion Laboratory, was satisfactory for colour and gelling quality.[11]

Māori people of Reef Point, Ahipara, and of Hokianga, Coromandel Peninsula, the Bay of Plenty, East Coast, and Hawkes Bay districts became the

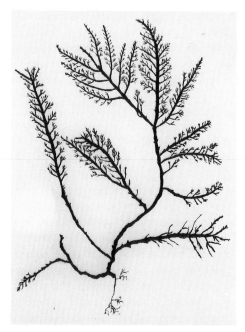

red alga (*Pterocladia lucida*) **Gelidiaceae**
Painting: V. Cassie

main collectors of *Pterocladia*. One shilling a pound was paid for dried weed, the industry was licensed to prevent competitive buying, and all collections were bought by Internal Marketing Division for resale to Davis Gelatine (N.Z.) Ltd — then the sole licensed manufacturer in New Zealand.

Production of agar began in July 1943, and in the first year the firm obtained fifteen tons of agar from about sixty tons of dried *Pterocladia*; an amount that was adequate for New Zealand needs. Agar is used to make a medium for the culture of micro-organisms, and as a gelling agent in canned meats, particularly sheep's tongues. It can stand autoclaving, and protects the meat against undesirable effects from metal containers, and from shaking during transport.[5,11,12,14]

Since trial gatherings of *Gelidium* and *Gracilaria* provided less dry weight of seaweed than the two species of *Pterocladia*, they were not continued. Studies of regeneration and growth rate by Lucy Moore indicated that northern beds of *Pterocladia* could be picked over annually, and this has been done without long-term damage.

By 1948, several tons of high grade agar were exported each year, the annual payout to seaweed collectors amounting to over £10,000. The amount of

Plate 9

kōtukutuku (*Fuchsia excorticata*) **Onagraceae** (p.103)
Painting: F. Osborne

Plate 10

kawakawa (*Macropiper excelsum*) **Piperaceae** (p.124)
Photo: J.E. Braggins

New Zealand cut-leaved nightshade (*Solanum laciniatum*) **Solanaceae** (p.165-6)
From a 1972 Russian series showing medicinal plants

Plate 11

poroporo (*Solanum aviculare*) **Solanaceae** (p.166)
Painting: F. Osborne

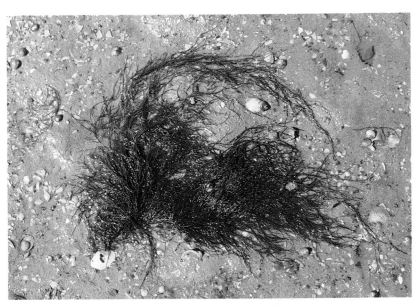

Gracilaria sp., 30 cm diam. (p.181)
Photo: W. Nelson

Plate 12

kōhia (*Tetrapathaea tetrandra*) **Passifloraceae** (p.103)
Painting: F. Osborne

Pterocladia collected averaged about 110 tons dry weight each year between 1950 and 1976, while agar extracted averaged about twenty-three tons; but local production has been falling since 1971, and there have been substantial imports of prepared agar since 1974.[2,12]

Until 1939, Japan was the main producer, but now many countries have their own agar industries, different agar types being recognized in the trade, with specific grades for various purposes.

Davis Gelatine (N.Z.) Ltd relinquished production of agar in New Zealand, and Coast Biologicals opened a new factory at Opotiki in 1979. Coast Biologicals now buys weed from 400 to 600 collectors round the coast; its major products are bacteriological grade agar, stabilizers, foliar and pasture fertilizers, and animal drenches. The company harvests some kelp seaweeds, for example *Eklonia*, for use in the fertilizers and animal health products. Kelp is also supplied to another company for manufacture into powders and granules for the health food market.[38]

13.4 *Gigartina* and Carrageen

Another seaweed import, carrageen, was threatened on the outbreak of war in 1939, and several local species of the red algal genus *Gigartina* were found to be suitable substitutes. Again the seaweed had to be picked from rocks or from heaps of weed thrown onto the shore during storms, rinsed in fresh water, dried, cleaned, and processed. The jelly extracted from local species of *Gigartina* was used successfully to clear the fines in beer-making and it has also been used in the local milk-custard industry.[12]

The name 'carrageen' was used in *Saunders' News-Letter and Daily Advertiser*, published in Dublin in 1829. Messrs Butler advertised Carrageen or Irish Moss, which they recommended as a 'Dietetic Remedy for Invalids' and as a substitute for 'Isinglas in Blamonge, Jellies, etc'. Efforts to trace a coastal locality named Carrageen have been unsuccessful and the origin of the name remains an Irish mystery.[13]

During 1972-5, some 300 collections of New Zealand species of *Gigartina* were prepared for standard carrageen analysis, the identification of the samples being made at the Botany Division, DSIR, Lincoln. Voucher specimens of each sample were preserved in the Herbarium. Analyses were made at the Applied Biochemistry Division, Palmerston North, and results showed that New Zealand has about twenty species of *Gigartina* from which carrageen can be extracted.[15-17]

Since 1975, a carrageen extraction industry based on New Zealand species of *Gigartina* has not gone ahead. For a while, Coast Biologicals of Opotiki processed another red alga, 'Zanzibar Weed' (*Eucheuma cottonii* var.) grown in Fiji, Tonga and Kiribati. The military coups in Fiji and the cost of shipping the seaweed to New Zealand ended the scheme. Now Coast Biologicals has a joint venture with a local firm for harvesting and processing an Indonesian seaweed.

13.5 *Macrocystis, Durvillaea* and Alginates

Moore (1941),[5] in a supplementary note to her study of kelp as a source of potash, pointed out that more than 16% of dry weight of *Macrocystis pyrifera* is alginic acid. She described an extraction process, suggested that alginic acid might be used to size *Phormium* fibre for textiles, and noted that sodium alginate is the starting point for seaweed fibres, fire-resistant transparent papers, and other new products. In 1948, she described Californian barge-harvesting of *M. pyrifera*, and listed many industrial uses of alginate, which was then an ingredient in half the ice-cream made in the US.[2]

In 1954, Betty Moss, King's College, Newcastle-upon-Tyne, and Margaret Naylor, Otago University, Dunedin, who collected and analysed plants of *Durvillaea antarctica* and *D. willana* from St Clair, Dunedin, found much higher levels of alginic acid; in some specimens of *D. antarctica* more than 40% of the dry weight of the lamina or blade was alginic acid.[18] In 1979, G.R. South, Memorial University, St Johns, Newfoundland, reported on a comparison of *D. antarctica*, of approximately known age from an experimental plot at Kaikoura, with adjacent and distant natural populations in the geographic range. He found higher alginate levels in *D. antarctica* than in *D. willana* of New Zealand and *D. potatorum* from Australia. He also discussed the importance of age fluctuations in alginate levels when determining optimum harvesting times.[19]

In 1979, Cameron Hay and G.R. South completed field experiments with *Durvillaea antarctica* and *D. willana*, the data having implications for the proposed commercial harvesting.[20] By 1990, the Chemistry Department, University of Otago, had found four common brown seaweeds to be 'rewarding in terms of yield of alginic acid'. These four were *Durvillaea antarctica, D. willana, Macrocystis pyrifera*, and *Marginariella boryana*. The structure of the alginate molecule is being studied to provide better knowledge of its properties.[21]

Many more alginate and related compounds such as sodium alginate are used in manufacturing than agar and carrageen — as emulsifiers, gelling agents, stabilizers, suspending and thickening agents, protective colloids, film and filament formers, and so on. The most important markets are the textile-printing and food industries. For food technology, alginates are especially useful as they can be used to hold together the ingredients of fabricated foods. As their use increases in New Zealand industry, there may be scope for a local alginate industry, but there is likely to be opposition to the harvesting of any brown alga from the southern coast, on the grounds that it might be prejudicial to young fish stocks.

13.6 *Gracilaria* and Agar and Pāua

In 1981, Karl Johnson, ARA, Mangere, reported to the Auckland Regional Authority Research Advisory Subcommittee on the 'aquaculture' of a red alga, *Gracilaria secundata* var. *flagellifera*, that grows most luxuriantly near the outlet of the Mangere sewage oxidation ponds, growth being promoted by nitrogen given off from the effluent in the form of ammonia. The Ministry of Agriculture and Fisheries granted licenses for the netting and harvesting of *Gracilaria* in the Manukau Harbour (Plate 11).[22]

The Plant Physiology Division, DSIR, Palmerston North, convened a committee to look at 'mariculture', the committee comprising representatives of other divisions of DSIR, MAF, ARA, the universities, and Coast Biologicals Ltd. In 1982, the committee recommended that research be initiated on the mariculture of *Gracilaria* and *Pterocladia* with a view to expanding the present agar industry.[23]

Although the quality of agar from *Gracilaria* is lower than that from *Pterocladia*, and the quantity is less, Ian Miller and Richard Furneaux, Chemistry Division, DSIR, Petone, developed a technique for upgrading *Gracilaria* agar to the standard required for bacteriology.[24]

The metabolism of agarophytes and carrageenophytes in response to factors such as light, temperature, CO_2 supply, and desiccation have been studied by Frank Dromgoole and students, Botany Department, University of Auckland. A *Gracilaria* and a *Gelidium* were chosen for initial work, the long-term aim being to provide a base for predicting optimal mariculture conditions.[25]

John Christeller and William Laing, Plant Physiology Division, DSIR, Palmerston North, studied physiological responses of *Gracilaria* from the Manukau Harbour; the aim was again to obtain information which might be of use in designing strategies for mariculture. Research was carried out in the Climate Laboratory, in specially designed tanks and natural sea-water, on the interactions between light, temperature, nitrogen nutrient levels, and growth rate and agar yield and quality.[26,27]

Margaret Gordon and students, Botany Department, Victoria University of Wellington, also initiated studies of unialgal cultures of *Gracilaria* from the Manukau Harbour in order to study methods of strain selection and micropropagation, regulation of development and production of spores.

At the ARA Manukau Sewage Purification Works, Mangere, Karl Johnson and MAF scientists found that young pāua between 5 and 25 mm long thrived in several cattle troughs containing *Gracilaria* grown in sewage effluent. Pāua

(*Haliotis iris*) is a univalve shellfish now in short supply as a seafood, and the possibility of 'reseeding' depleted beds with young raised on *Gracilaria* has been the subject of a number of studies.[28]

At the Mahanga Bay Shellfish hatchery of MAF near Wellington, work on *Gracilaria* as a food for cultured pāua was funded by a Fisheries Research Division Research Contract with Victoria University of Wellington. The *Gracilaria* used was sent from Auckland, and studied to determine best growth conditions to produce pāua food and agar. The Fishing Industry Board, Fisheries Research Division of MAF, and Victoria University of Wellington arranged preliminary trials of a system of spray culture of *Gracilaria*, and reviewed pond culture of the seaweed overseas.

Young pāua are fed on a film of epiphytic diatoms until they are large enough to eat seaweed. A red alga, *Audouinella*, epiphytic on *Gracilaria*, seemed to provide the pāua with a surge of growth. Changing the pāua from *Gracilaria* to a diet of *Macrocystis* and back again also seemed to result in surges of growth.[29]

In November 1986, Thomas Mumford, State Phycologist and Project Leader for Seaweed Farming, Washington State, Seattle, US, gave seminars in New Zealand at the invitation of the Fishing Industry Board. At the Auckland seminar, Mumford dismissed *Gracilaria* seaweed farming as unlikely to be economic here. Prices are volatile and low at present. *Eucheuma* farming in developing tropical countries is much more likely to be successful. He urged New Zealand scientists and businessmen to aim for a high-quality product at the upper end of the seaweed market. He also emphasized that adequate research must precede the launching of commercial operations.[30]

At the University of Otago in 1987, Anita Pillai, a Ph.D. student in the Botany Department, began an ecophysiological study of *Gracilaria* to determine growth rates, biomass and regeneration, especially from 'maricultural plants'.[31] About the same time (1987), Wendy Nelson, National Museum, Wellington, completed the revision of the New Zealand species of *Gracilaria*. The Manukau population, known as *G. secundata* var. and f. *flagellifera*, as *G.* sp. nov., and as *G. sordida*, is now *G. chilensis*.[32,39]

Since the 1986 Auckland seminar, it seems that the best prospect for *Gracilaria* will be as food for young pāua. However, there are problems to be solved; for example, staff at the University of Auckland Leigh Marine Research Station found that hatchery-reared pāua from Mahanga Bay, which were put into the marine reserve at Goat Island, opposite the Leigh Station, in the Hauraki Gulf, did not survive storms. Possibly they were put in the wrong place, or they may have been stressed by too much handling. In 1989, the Mahanga Bay Hatchery began raising young pāua on *Gracilaria* to replenish depleted stocks on the Chatham Islands. Young pāua were brought from the Islands for that scheme, as pāua there are of a different kind.

13.7 *Porphyra* and Nori

Wendy Nelson, National Museum, Wellington, has begun a revision of the New Zealand species of *Porphyra*, collaborating with Kathleen Cole, University of British Columbia, on chromosomes, with Margaret Gordon and Bill Brostoff, Victoria University of Wellington, on electrophoresis, and with Charles O'Kelly, Massey University of Manawatu, on life histories. O'Kelly announced plans to study different Māori perceptions of karengo 'on a partnership basis in both philosophical and economic terms'. The C. Alma Baker Research Fund granted funds for O'Kelly's work.[33]

S.F. Friedlander, L.D. Melton, and M.T. Brown of the Food Science and Botany Departments, University of Otago, investigated reputed high levels of ascorbic acid in *Porphyra columbina* and *P. subtumens* of New Zealand. They found that actual levels are dependent upon species, location of the plants, time of collection, and length of storage. Highest levels were found when the plants were in a vigorous growth phase, tender and richly pigmented. Further research is needed on storage to maintain high ascorbic acid levels. Meantime, a pilot-scale 'nori' farm has been established with support from southern businessmen and the assistance of the University of Otago Seaweed Biology Research Group.[34]

The Japanese people use a number of species of *Porphyra* for food, consuming more than 20,000 tons or 7.7 billion sheets of 'Asakusa-nori' each year; the dried seaweed sheets are fragrant when toasted, and are put in other dishes or used as a wrap around rice foods. Already, work in Otago suggests that New Zealand can produce a nori of acceptable quality, and hope has been expressed that a local fast-food chain will follow the successful US practice of marketing rice foods wrapped in nori.[35]

13.8 *Undaria* and Wakame

At the Marine Sciences Conference held at Dunedin in August 1987 to coincide with the reopening of the Portobello Marine Laboratory, Cameron Hay, Division of Marine and Freshwater Science, DSIR, Wellington, reported the discovery in Wellington Harbour of luxuriant beds of the Japanese seaweed, *Undaria pinnatifida*. This brown alga has been introduced by accident, and is now well established.

Undaria or 'wakame' is a popular Japanese food; about 100,000 tons of the dried and processed seaweed are consumed each year. It is used in soups, marinated dishes, roasted, and candied with sugar, and is an essential item in the traditional Japanese diet. The prospects for harvesting and marketing the Wellington Harbour plants were described as promising.

13.9 Current Research

Studies of *Gracilaria* and *Porphyra* by Margaret Gordon and students, Botany Department, Victoria University of Wellington, have been mentioned. Interesting hormone effects were found during examination of the regeneration of 1 mm segments of *Gracilaria* and *Pterocladia*.

M.H.G. Munro, D.J.W. Blunt, and co-workers, Chemistry Department, University of Canterbury, Christchurch, have continued their ongoing project which involves antiviral, antimicrobial and toxicological assay of extracts from marine animals and plants. The identification of the seaweeds is carried out at the Botany Institute, Land Resources, where herbarium sheets are kept. The failure of Sea Pharm in 1988, loss of the US National Cancer Institute contract, and the forced termination of the employment of a number of skilled divers caused some problems. However, a five year contract has been negotiated with Pharma Mar SA (Spain), an active programme of collecting has continued, and the chemistry effort is maintained.[36]

During 1986, an Aquaculture Research Centre was set up at the University of Otago, under the directorship of Professor Margaret Loutit. The Centre comprised an informal affiliation of staff from several disciplines, the core being Botany, Chemistry, Human Nutrition, Microbiology, Zoology, and the rebuilt and expanded Portobello Marine Laboratories. About thirty staff in ten departments have been involved in activities of the Centre. In March 1990, after Philip Mladenov, newly appointed Professor of Marine Science, took over the directorship from Professor Loutit, the scope of the Aquaculture Centre was redefined.

In June 1989, the Government announced that the 'bureaucratic restrictions' on marine and freshwater farming would be removed as soon as possible to provide for new industries, more employment, and increased exports. Thomas Mumford's advice regarding the need for thorough research before starting a new venture was not mentioned.

A comment by Brian Coffey, formerly with MAF and now a consultant at Hamilton, seems relevant. He suggested that some of the problems encountered in fish and shellfish farming arose from the surplus nutrients produced at the farm sites. He wondered whether seaweed farming alongside the caged fish or tethered shellfish would 'mop up' any surplus nutrients and provide a healthier environment.[37]

1. Colenso, W. (1869a, 2nd ed. 1875) 1 *Trans. N.Z. Inst.* 233.
2. Moore, L.B. (1948) 2(13) *Post Primary School Bulletin* 257. Reprinted as part of Moore, L.B., Adams, N.M. *Plants of the New Zealand Coast* Auckland (Longman Paul) 1963 113p.
3. Dieffenbach, E. *Travels in New Zealand* London (J. Murray) 1843, 2 vols. 431 396p. Capper Press reprint 1974.

4. Aston, B.C. (1916) 13 *N.Z.J. Agric.* 446.
5. Moore, L.B. *The Economic Importance of Seaweeds* 1941, reprinted 1966. DSIR Bull 85, 40p.
6. Rapson, A.M., Moore, L.B., Elliott, I.L. (1942) 23B *N.Z.J. Sci. Tech.* 149.
7. Grimmett, R.E.P., Elliott, A.G. (1940) 61 *N.Z.J. Agric.* 167.
8. Little, E.C.S. 1948 M.Sc. Thesis, University of Auckland.
9. Francki, R.I.B. (1960) 12 *Plant & Soil* 297, 311.
10. Francki, R.I.B. (1964) 20 *Plant & Soil* 65.
11. Moore, L.B. (1944) 25B *N.Z.J. Sci. Tech.* 183.
12. Chapman, V.J., Chapman, D.J. *Seaweeds and their Uses* London (Chapman & Hall) 1980, 3rd ed. 334p.
13. Mitchell, M.E., Guiry, M.D. (1983) 9 *J. Ethnopharmacology* 347.
14. Moore, L.B. (1942) 23B *N.Z.J. Sci. Tech.* 170.
15. Parsons, M.J., Pickmere, S.E., Bailey, R.W. (1977) 15 *N.Z.J. Bot.* 589.
16. Pickmere, S.E, Parsons, M.J., Bailey, R.W. (1973) 12 *Phytochemistry* 2441.
17. Pickmere, S.E, Parsons, M.J., Bailey, R.W. (1975) 18 *N.Z.J. Sci.* 585.
18. Moss, B., Naylor, M. (1954) 81 *Trans. Roy. Soc. N.Z.* 473.
19. South, G.R. (1979) *Proc. 9th. Inst. Seaweed Symposium* Santa Barbara, California, US, 1977, p. 133.
20. Hay, C.H., South, G.R. (1979) 22 *Botanica Marina* 431.
21. Parikhar, R. (1990) 2 *University of Otago Marine Science* Bulletin 3.
22. Johnson, K.A. 1981 'Gracilaria Aquaculture Project'. Final Report to Auckland Regional Authority Advisory Subcommittee, 23p.
23. Christeller, J.T. et al. 1983 'The Potential for Mariculture of Seaweeds in New Zealand'. N.Z. DSIR Plant Physiology Division, Palmerston North, Tech. Report 18, 21p.
24. Miller, I.J., Furneaux, R.H. (1982) 25 *N.Z.J. Sci. Tech.* 15.
25. (1986) 29 *N.Z. Marine Sciences Society Newsletter* 33.
26. Laing, W.A., Christeller, J.T., Terzaghi, B.E. (1989) 32 *Botanica Marina* 439.
27. Christeller, J.T., Laing, W.A. (1989) 32 *Botanica Marina* 447.
28. *New Zealand Herald* 28 January 1984 'Paua thrive on Sewage Weed'.
29. Gordon, Margaret. Victoria University of Wellington. Letters of 15 and 25 September 1989.
30. Mumford, Thomas. State Phycologist, Washington State, US public seminars, November 1986.
31. Pillai, A. (1987) 7 *N.Z. Bot. Soc. Newsletter* 6.
32. Nelson, W. (1987) 25 *N.Z.J. Bot.* 87.
33. *Waikato News* 21 December 1989, p. 17: C. O'Kelly on karengo.
34. Friedlander, S.F., Melton, L.D., Brown, M.T. (1989) 37 *Japan J. Phycol.* 295.
35. Arasaki, S., Arasaki, T. *Low Calorie, High Nutrition Vegetables From The Sea* Tokyo (Japan Pubs Inc) 1983 196p.
36. (1989) 32 *N.Z. Marine Sciences Society Newsletter* 28.
37. Coffey, B.T. Consultant, Hamilton, pers. commun.
38. Coast Biologicals Ltd, P.O. Box 58-103, East Tamaki, letter of 23 November 1990.
39. Dr Wendy Nelson, letter of 8 April 1991 to R.C. Cooper.

14

Marine and Freshwater Microalgae

Microalgae may be aquatic or terrestrial, and the aquatic kinds may be marine or freshwater, but a small number of species are found in both freshwater and the sea. Microalgae are usually single-celled, minute, and are seldom seen save perhaps as an area of coloured water in the sea, a froth or stain on the sand at high water mark, a scum on a pond, or a coloured felt below a thermal hot pool. Planktonic species float in the water, benthic species are attached to the bottom, while terrestrial species lead a sedentary life attached to various surfaces.

Most species of microalgae found in New Zealand also occur in countries overseas and we have few endemic species; consequently, research on the taxonomy, ecology, uses, and so on, of foreign microalgae may have immediate relevance here and it is necessary to keep watch on the enormous world literature.

At present, there are few local guides to New Zealand species, although a start has been made: F.J. Taylor, Leigh Laboratory, University of Auckland, prepared annotated checklists of marine microalgae in 1970 and 1974.[1-3] The late R.D. Wood, University of Rhode Island, and Ruth Mason, Botany Division, DSIR, Lincoln, revised the fifteen New Zealand species of Characeae in 1977;[4] Vivienne Cassie, Botany Division, DSIR, Mount Albert, finished an annotated checklist of freshwater algae (excluding diatoms and charophytes) and another list of freshwater diatoms in 1984;[5,6] and Hannah Croasdale, Dartmouth College, New Hampshire, together with Elizabeth Flint, Botany Division, DSIR, Lincoln, completed Volume 1 of a three-volume Desmid Flora in 1986, and Volume 2 in 1988.[7,8] The following year (1989) Vivienne Cassie completed a two-volume work on the ecology and taxonomy of 220 species of the thermal microalgae of New Zealand[9-11]

In the present state of knowledge, numbers are low and approximate. The total number of species of marine microalgae in New Zealand is estimated at 700, including 450 species of diatoms; the number of species of New Zealand freshwater microalgae is 2076, including 548 species of diatoms. The three-volume Desmid Flora, when complete, will contain descriptions of about 690 species.

14.1 Phytoplankton and Productivity

The plankton of the seas and inland waters around New Zealand consist of a large population of microscopic plants and animals. The plants, containing chlorophyll and performing photosynthesis, provide organic products which are the bases of both marine and freshwater food chains.

Recent studies of primary productivity in coastal waters include those of Janet Bradford, Hoe Chang, and colleagues, Division of Water Science, DSIR, Wellington, on the primary productivity and zooplankton distribution off Westland, and factors controlling summer phytoplankton in greater Cook Strait;[12,13] Lincoln McKenzie and P. Gillespie, Cawthron Institute, Nelson, on bacterioplankton biomass and heterotrophic activity off Westland; P. Gillespie (ed.) on nutrient processing and biomass production in New Zealand estuaries;[14] and Webber Booth, Botany Department, University of Auckland, on the distribution and abundance of phytoplankton in Hauraki Gulf (for the MAF Hauraki Gulf Snapper Recruitment Programme).

Two interesting items came from Booth's research: the first was the demonstration that epiphytic diatoms form an important part of the combined photosynthetic rate of the algal host and epiphyte; the second was the recovery and description of phytoplankton that had previously passed undetected.[15,16]

Working with M. Sondergaard, a visiting scientist from the Botanical Institute, University of Aarhus, Denmark, Booth obtained picoplankton from the Hauraki Gulf for the first time. These are cells which are capable of passing through a filter with 2μm pores, but not through one with 0.2μm pores. Overseas, studies (e.g. Fogg, 1986) have shown that cells of this size are more efficient than larger cells in absorbing radiant energy, and their role in recycling nutrient elements is of great importance in the marine ecosystem.[17] In 1988 Julie Hall of Massey University, Palmerston North, and Warwick Vincent of the DSIR Taupo Laboratory studied the picoplankton of the upwelling sea water off Westland during midwinter, when several commercially important kinds of fish migrate there to feed. They found that picoplankton were the major contributors to the algal biomass and food web dynamics of the West Coast fishery.[72]

Over many years there have been accounts in the New Zealand press of the decline in toheroa numbers on Ninety Mile and other beaches. Stories of reduction in fish numbers are common. Tales of seabirds dying of starvation are less common, the latest coming from the Bay of Plenty in November 1990 and reporting the death of hundreds of shearwaters.[65] A decline in marine phytoplankton productivity might be responsible for the fall in shellfish, fish, and bird numbers; however there is almost no evidence to support the hypothesis.

Studies of nutrients, phytoplankton species and biomass are available for only a few localities in New Zealand coastal waters, for example off Westland (1979),[13] Goat Island, Leigh (1981 and 1985)[66,67] Cook Strait (1986),[12] Tasman Bay (1986)[68] and Kenepuru Sound (1986).[69] For most of our coastal waters there are no baselines against which changes in productivity might be measured.

Phytoplankton biomass in Goat Island Bay, measured in 1966-9 and again in 1981-2 showed no apparent relation to increased nitrate present. However, significant correlations with radiation received were suggested.[66,67]

Overseas, there are several reports of phytoplankton productivity being far below what it could be, considering the levels of phosphates and nitrates in the seawater sampled. This has led to the suggestion that more research is needed, perhaps on a world basis.[70] A shortage of trained people and a lack of funds seems likely to prevent New Zealand from making much progress in this field.

For the history of research on the freshwater phytoplankton of New Zealand, reference should be made to *New Zealand Lakes*, 1975, edited by V.H. Jolly and J.M.A. Brown, to a bibliography of the freshwater algae of New Zealand 1849-1980, prepared by Vivienne Cassie (1980), and to *Inland Waters of New Zealand*, 1987, edited by A.B. Viner.[18-20]

In the 1975 *Lakes* book, Elizabeth Flint, then Ecology Division, DSIR, Vivienne Cassie, then Botany Department, University of Auckland, and Caroline Burns, Zoology Department, University of Otago, provided lists of phytoplankton species for a total of seventy-five New Zealand lakes, related the species to water quality, and classified the lakes as eutrophic, mesotrophic, or oligotrophic[21,22,24] In the 1987 *Inland Waters* book, A.B. Viner and E. White discussed several factors affecting phytoplankton growth, using the records compiled by Flint, Cassie, and Burns (1975), Cassie (1978), and other authors.[21,24] In both books, the reader is scarcely made aware of the vital role of phytoplankton in freshwater as primary producers of food for the smallest animals in the food chain.

John Stockner of Fisheries and Oceans, Canada, worked with Carolyn Burns at the University of Otago in October 1989 on new techniques for the enumeration of freshwater picoplankton. The introduction of these techniques opens a new field of plant life for study. Carolyn Burns has completed a survey of the picoplankton in six southern lakes of different trophic status. ('Could the refreshing taste of Lake Manapouri's water be due to 0.2 million picoplankton per ml?').[71]

14.2 Diatoms and Diatomite

Diatoms are a group of unicellular microalgae with yellow-brown pigments. They are among the most numerous member of the phytoplankton. In the aquatic system, which covers 70% of the global surface, diatoms are thought to be the major users of carbon dioxide and to provide a large percentage of the world's atmospheric oxygen.

Knowledge of the ecology and taxonomy of diatoms is still incomplete and the study of the group assumes greater significance at a time when global ecological considerations such as pollution and global warming are causing concern.[25]

They are also important in benthic communities; for example, Vaughan Wilkinson, Botany Department, University of Auckland, in an M.Sc. thesis (1981), showed that chlorophyll concentrations and primary productivity from

A centric diatom (*Cyclotella stelligera*) from Puarenga stream, Whakarewarewa
Photo: W. Booth
Original magnification x 2840

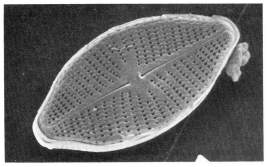

A pennate diatom (*Navicula salinarum*) from a hot pool at Whakarewarewa
Photo: W. Booth
Original magnification x 4400

the 'microphytobenthos' (diatoms living on and just below the mud surface), markedly exceeded those from the phytoplankton in the waters of the Manukau Harbour. He suggested that monitoring diatom populations living on the mud surface would indicate the changing capacity of the harbour to assimilate wastes from sources such as the Mangere sewage treatment works.[26]

When diatoms die, their siliceous cases persist, and diatomite may form as a layer at the bottom. From the famous marine diatomite deposit at Oamaru, about 800 fossil diatom species of Eocene age have been described in approximately 125 scientific papers. In 1989, T.V. Desikachary and P.M. Sreelatha revised 647 Oamaru diatoms.[73] The largest freshwater diatomite deposit in New Zealand, at Middlemarch, Otago, is of Upper Miocene age, and contains about 50,000 tons of white diatomite in pure (unialgal) beds up to forty metres thick, along with about five million tons of black diatomaceous earth.[27,28]

There are over sixty localities in New Zealand where diatomite has been found, usually where volcanic lava or ash ponded a still body of water at some time in the past. No New Zealand deposits are worked at present, and some have been destroyed, for example, those at Kamo (Whangarei), and at Morningside (Auckland).[29] Diatomite is chemically inert, porous, insoluble, and each shell is rigid in structure; consequently, it is valued for its high heat, fire and electrical resistance and has a variety of uses. Diatomite has been mined in New Zealand for insulation, filtration, and pozzolana in concrete, and it is possible that local deposits will be used again, particularly since diatomaceous earth has been found to be one of the most effective filters for the removal of polio and similar viruses from wastewaters.

14.3 Marine Algal Blooms

Blooms of planktonic algae occur when conditions of nutrition enable them to reproduce very rapidly. Most blooms are not toxic — that is to say, they do not cause the death of other organisms; but some marine blooms of plants and animals are responsible for allergies such as 'swimmers' itch', 'Doggerbank itch', and 'Tamandare fever'.[30]

The extent of seawater discoloration in New Zealand coastal waters has been studied for some years, and the phenomenon, often called a 'red tide', is common, especially in Cook Strait. Apparently these blooms of protozoa, tunicates, algae, and other organisms are triggered by an increase in phosphates and other minerals due to the upwelling and mixing of seawater.

Vivienne Cassie (1981)[31] described the 'brown waters' seen off Westport in the Karamea Bight during March 1976, and again in March 1978. Hoe Chang (1988)[32] recorded a similar bloom in the same locality in the later summer of 1982, noting that it covered an area of about 20,000 square kilometres. The dinoflagellate *Prorocentrum micans*, which caused the three blooms, is sometimes toxic; however, there were no reports of fish or shellfish kills.

Hoe Chang (1983)[33] discussed the occurrence of net-clogging slime in Tasman Bay during the spring of 1981, which reduced fish catches and severely affected the livelihood of fishermen. The phenomenon is a recurring feature in Tasman Bay, having been first noted in the 1860s. The mucilage-producing microalga was identified by Chang as a colonial form of *Phaeocystis pouchetii* (Prymnesiophyceae). *Gyrodinium aureolum*, a toxic dinoflagellate, was identified from the Tasman Bay slime by Vivienne Cassie (unpublished, 1985). In January 1985, a similar slime clogged nets and coated long lines in parts of the Hauraki Gulf.

Until the 1980s, none of these phenomena in our coastal waters were known to have produced toxic effects, but from late December 1982 to mid January 1983 there were massive fish and shellfish kills off the eastern coast of the North Island. F.J. Taylor and colleagues (1985) had studied annual diatom blooms at Leigh Marine Research Laboratory for several years, and found in January 1983 that the diatom *Cerataulina pelagica* was sixty-five times more numerous than usual, forming a dense blanket of dead cells up to 1.5 metres deep. They extrapolated the Leigh results to other localities, and attributed the death of benthic shellfish at Bream Bay, on the Coromandel Coast, and in the Bay of Plenty, to anoxia induced by bacterial decay of the diatoms, and the death of bony fish caught on long lines to anoxia and to the clogging of the gills with mucilage produced by the diatom.[34] However, Chang and Ryan (1985)[35] described and named a new species of *Prymnesium*, *P. calathiferum*, from among the dominant phytoplankters at the time of the shellfish and fish kills. Species of the genus *Prymnesium*, and in particular *P. parvum*, produce potent toxins and cause epidemic mortalities of shellfish and fish overseas. A unialgal culture of the new species was found to be toxic to fish, and Chang and Ryan concluded that it was probably responsible for the 1982-3 fish and shellfish kills off the eastern coast of the North Island as the level of oxygen in the water column was not low enough to cause mortalities of marine animals.

Further evidence of toxicity came from a blood-red tide in the upper Whangarei Harbour three weeks after the first shellfish deaths were reported from Bream Bay. From this red tide, Vivienne Cassie isolated a probable new species of the toxic dinoflagellate genus *Alexandrium* (formerly *Protogonyaulax*). She forwarded a sample to Dr F.J.R. Taylor, Department of Oceanography and Botany, University of British Columbia, Vancouver, who cultured the species and studied it for toxin production.[36]

The only conclusion possible from this evidence is that further studies are needed regarding the life cycles of toxic microalgae inhabiting New Zealand coastal waters. An international workshop on toxic dinoflagellates was held at Cronulla, near Sydney, in June 1984,[37] and the techniques learnt by two New

Zealand scientists who attended have been applied to a joint Fisheries/Oceanographic Institute project to look for benthic cysts of toxic dinoflagellates in sediments from New Zealand shellfish growing areas. In the same field, Cawthron Institute, Nelson began two projects in 1986; first to determine the distribution of toxin-producing phytoplankton in the Marlborough Sounds, and second, to develop and standardize chemical procedures for extracting and assaying mussel flesh for phytotoxin levels.[38,39]

In early January 1989 there was a toxic bloom in Big Glory Bay, Stewart Island, where salmon were raised in cages. The alga was about twenty microns long, bean-shaped, with two flagella and, at the height of the bloom, was present at the rate of seventy million per litre. The sea was coloured tea-brown, and the loss of salmon has been estimated at various figures ranging from $78,000 to millions. The alga was identified as *Heterosigma akashiwo*, which has been linked to the death of caged fish in the Inland Sea of Japan, in salmon farms in Canada in 1985 and 1986, and in Chile in 1988.[40,41]

14.4 Freshwater Algal Blooms

Algal blooms on lakes and ponds may be fed by the inflow of sewage effluent, or fertilizer from adjoining farm land, or even the droppings of ducks or swans. If the bloom is of blue-green algae (some are of green algae), honey bees, fish, birds, dogs, stock, and man may be poisoned. Elizabeth Flint, then freshwater phycologist, University of Canterbury, Christchurch, listed four toxic species of blue-green algae and associated bacteria in 1966, described the symptoms of poisoning, advised how to build farm ponds to avoid algal blooms, and recommended copper sulfate drags to kill the algae.[42]

Henry Connor, Botany Division, DSIR, Lincoln, included Flint's list, now raised to six species, in his work on poisonous plants in New Zealand, named places where algal blooms have occurred, discussed the many suspected and few proven cases of stock losses, and mentioned the suspicion that diarrhoea in man may be caused by a blue-green algal toxin in drinking water.[43]

The colourless unicellular alga *Prototheca wickerhamii* can also be pathogenic to humans and animals. About 1974, two cases of 'protothecosis' in humans were found in Auckland, and the alga was isolated from the blood of one patient and the ear discharge of another. *Prototheca* has been responsible overseas for abscesses and lesions in humans, and massive lesions causing death in cattle. The cells, which are colourless and sometimes mistaken for yeasts, have been found in soil, wastewater ponds, lavatory drains, slime on plants, and the slimy water of flower vases should probably be treated with caution.[44,45]

During the 1982-3 summer B.J.F. Biggs, Hydrology Centre, Ministry of Works, Christchurch, circularized twenty regional Water Boards and obtained returns showing that over 100 waterways throughout New Zealand are subject to some degree of algal blooming, especially in summer. The green alga *Spirogyra* is most common, but other green algae, *Cladophora* and *Rhizoclonium*, form conspicuous growths. A number of Boards noted a relationship between the proliferation of algae and the intensity of agricultural development in the catchments. Major algal blooms were observed to reduce the value of waterways for swimming and fishing, blocked irrigation channels, and led to a degradation of water quality.[46]

In 1985, B.T. Coffey and S. Miller, Ruakura Agricultural Research Centre, MAF, Hamilton, found that 73% of farm troughs and dams surveyed between Hamilton and Kaitaia contained blue-green algae. The only toxic alga, a blue-green species, *Microcystis aeruginosa* (also known as *Anacystis cyanea*) was present in 16% of troughs and dams. Connor (1977) recorded that sheep, cattle and pigs are susceptible to the toxin which is released by ageing cells. Taste and odour algae were also present in 16% of troughs and dams. No attempt was made to estimate the loss to farmers caused by these algal blooms.[47]

Finally, mention must be made of the 1985-6 algal bloom at Lake Omapere. The lake is 1220 ha in area, about 1.5 m deep, and has a warm (thermal) spring in one corner. In November 1985, the toxic algal bloom killed birds at the lake and forced the Borough of Kaikohe to draw its water from other sources. Algae washed out of the lake down the Utakura Stream, rendering its water undrinkable, and fouled the upper reaches of the Hokianga Harbour, making both fish and shellfish inedible. Options investigated by the Northland Catchment Commission, Whangarei, included control of swan numbers, introduction of algae-eating silver carp, and buffer strips. Again, the culprit algae was *Microcystis aeruginosa*, with two other toxic blue-greens, *Anabaina circinalis* and *A. flos-aquae*. In 1988 the Bay of Islands Electric Power Board decided to study the possibility of raising the lake level to flush out the algae and to provide water for a power station. An alternative scheme was also suggested about this time, to raise the lake level to provide water for an irrigation scheme for the Waimate North area.

Apparently, the buffer strips envisaged by the Northland Catchment Commission task force were cut-off ring drains around the lake to stop phosphates entering it, and/or an area of artificial wetland at the point where the lake flows out into the Utakura River. There are problems with the boundaries of some properties owing to erosion. Many interests are involved, ranging from Māori owners who wish to preserve their eel fishing, to local farmers who want to water their stock. For some reason, a buffer strip of native plants, Radiata pine, or other vegetation, around the edge of the lake does not seem to be an option for consideration. As the blooms are recurring, and water supplies in Northland are limited, more will probably be heard of Lake Omapere algal blooms in the years to come.[48,49]

14.5 Microalgae, Sewage, and Water Quality

Since 1950, numerous ponds and lagoons have been built throughout the world as an essential part of sewage purification works; for example, near Auckland, there are extensive ponds and lagoons at Mangere as part of the Manukau Sewage Purification Works, further ponds at Rosedale Road, near Albany, as part of the North Shore Purification Works, smaller ponds at Wellsford, Warkworth, Snells Beach, Waiwera, Orewa, Helensville, Clarks Beach, St Stephens School at Bombay, Wesley College at Paerata, the Paerata Dairy Company, and Taukau. Truly our engineers, aided by officers of the Health Department, have done well out of ponds!

Some of these ponds receive raw sewage with no primary treatment other than the coarse screening-out of larger debris, some are sludge lagoons receiving anaerobically digested primary sludge or activated secondary sludge underwater, some are oxidation ponds where secondary sewage (wastewater from which large debris has been removed by screening and sedimentation) is treated, and some are 'polishing ponds' where further treatment is given.

In 1970, Arthur Haughey, ARA, Mangere, described how *Euglena acus*, the dominant euglenoid flagellate, plays an important role in maintaining an oxygenated layer of water above the material in the sludge lagoons, and thus prevents an odour nuisance in the district.[52]

Between 1961 and 1968, Haughey collected, identified, described and illustrated fifty species of freshwater microalgae found in sludge lagoons, oxidation and polishing ponds near Auckland. From a second pond survey, made between 1977 and 1982, Vivienne Cassie identified, keyed and illustrated a total of seventy species. Thus members of the flora can now be given a name.[50,51,53,54]

Oxygen can be introduced by mechanical means to stabilize finely divided and dissolved wastes, but it is far cheaper to use the microalgae living at or near the surface of the ponds. These algae take energy from sunlight, food from air, water and wastes, and produce oxygen as a by-product of photosynthesis. The dissolved oxygen sustains a rich bacterial flora and a similar protozoan fauna, which live and feed in lower layers of the pond.

The aim of oxidation pond management is to maintain a surface layer of microalgae, rich in species, but with seasonal changes problems can arise; for example, if the water temperature rises above 23°C, and there are no high winds to disturb the water surface, a bloom of a blue-green species may occur. Usually, such a bloom consists of a single species, and when it dies nauseating odours may be released.

At present, about 60% of New Zealand's sewage effluent is discharged into coastal waters, and the balance is released into lakes, rivers, and even sand dunes, sprayed onto land, or used to irrigate Radiata pines. Micronutrients,

Plate 13

pūriri (*Vitex lucens*) **Verbenaceae** (p.134)
Painting: E. Blumhardt

Plate 14

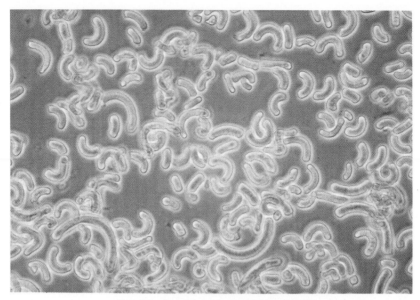

Spirulina from a Japanese sample (p.198)
Photo: V. Cassie
Original magnification x 640

vegetable caterpillar (*Cordyceps robertsii*) (p.206)
Haast Pass. Photo: E. McKenzie

Plate 15

NEW ZEALAND PLANT DYES (p.133)

Mordants 1%

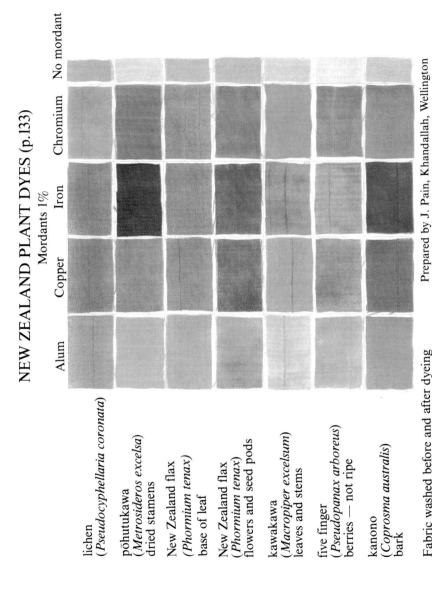

Columns: Alum, Copper, Iron, Chromium, No mordant

Rows:
- lichen (*Pseudocyphellaria coronata*)
- pōhutukawa (*Metrosideros excelsa*) dried stamens
- New Zealand flax (*Phormium tenax*) base of leaf
- New Zealand flax (*Phormium tenax*) flowers and seed pods
- kawakawa (*Macropiper excelsum*) leaves and stems
- five finger (*Pseudopanax arboreus*) berries — not ripe
- kanono (*Coprosma australis*) bark

Prepared by J. Pain, Khandallah, Wellington

Fabric washed before and after dyeing

Plate 16

mairehau (*Phebalium nudum*) **Rutaceae** (p.153)
Painting: E. Blumhardt

such as phosphorus, nitrogen and organic carbon, enrich the waters into which the effluent is discharged, with consequent effects on the plants and animals around the outfall.

With increasing eutrophication, due in part sometimes, but not always, to sewage effluent, the maintenance of good water quality has become much more important. In 1979, Carolyn Burns surveyed fifteen possible techniques for the biological assessment of water quality, and concluded that further testing and standardization of techniques and criteria were necessary. Chemical and bacterial assays are commonly used to determine water quality, but specific analyses of algae and animals present may provide a better long term assessment.[55]

The work of Margaret Loutit and co-workers in Dunedin on the persistence of human enteroviruses in waste water, and in the environment around and downstream from sewage effluent outfalls, is discussed in the section in Chapter 15 on viruses, but it is mentioned here as it endorses the suggestion that techniques for assessing water quality could be improved.

The usual result of eutrophication is that the number of algal species present falls, and the number of individuals of the surviving species rises; consequently algal species present and algal species absent, particularly diatoms, are indicators of the degree of organic pollution in a body of water. In Vivienne Cassie's contribution to the 1979 survey, however, she concluded with the comment that a wider knowledge of the taxonomy of benthic diatoms was imperative before this group of algae could be used with confidence in water quality assessment.[44]

Some eleven years later, there is still a dearth of benthic algal identifications to species level in many New Zealand water quality reports (except in 'The 100 Rivers Survey' by B.J.F. Biggs and co-workers of the Hydrology Laboratory, DSIR, Christchurch).

14.6 Microalgae of the Future

Botryococcus braunii is a common freshwater green alga which has been studied overseas as a source of microalgal oils and fats. The volatile content (equivalent to a loss on ignition of 99%) is consistent with the very high amount of hydrocarbon known to be present in the alga. The mean daily export of the alga from Lake Taupo is more than 230 kg of dry weight. A small but efficient hydrocarbon extraction plant at the outlet of Lake Taupo might be economically viable and assist in reducing problems of biological productivity downstream.[56]

Spirulina, a synonym of *Oscillatoria*, is widely known as a genus of blue-green algae with highly spiralized trichomes (hairlike rows of cells curled in spirals). Species of *Spirulina* occur in New Zealand; the well known *S.*

platensis has been reported from Lake Rotoiti and the Mangere oxidation ponds (Plate 14).[53]

There is a large-scale plant at Lake Texcoco, Mexico, where *S. platensis* is harvested from a giant solar evaporator, homogenized, pasteurized, and spray-dried. The product is rich in protein; in some tests protein content was 70% of dried weight. In New Zealand, tablets are made from imported dried *Spirulina*, and sold in health-food shops as a substitute for meat, and as a pre-meal supplement to reduce food intake. Supplies have come from Mexico, Israel, Taiwan, and Japan, where the technology is well-known.

In future, New Zealand acquaculturists might find that growing *Spirulina* on wastewater is an attractive possibility. Wastewater nutrients could be recycled and the protein obtained could be used as a stock food, as a human food supplement, as a material for hydrogen production, and as a source of phycocyanin, a pigment used for food colouring, fermentation, cosmetics, and for immunofluorescent assays.[57]

Spirulina is a source of dietary gamma-linolenic acid; however evening primrose oil is cheaper, and even better sources are likely to become available.[58]

In 1978, studies were begun at the Roche Research Institute of Marine Pharmacology, Dee Why, NSW, on the production of beta-carotene, glycerol and feed protein from *Dunaliella salina*, a unicellular green alga that sometimes colours salt-ponds pink. When Roche withdrew in 1981, research was continued by Drs Lesley and Michael Borowitzka and colleagues, assisted by Commonwealth Government grants. During 1985-6, Western Biotechnology Ltd, Perth, was launched to purchase the research, run a 25 ha ponds system, and market the product. Dr Lesley Borowitzka was appointed director of production and marketing.

The company's main product is beta-carotene, a yellow-red plant pigment, extracted from *Dunaliella* by a solvent-free method developed during the years of research. By 1986, the company was selling various solutions of beta-carotene worldwide for food colouring, and was looking at marketing it in concentrated form as a dietary supplement, supplying vitamin A and apparently playing some part in cancer prevention and remission. The company was also studying evidence that beta-carotene improved cattle and sheep fertility.[59-61] Beta-carotene is now a popular pill in health-food shops.

In 1986, Hoe Chang and colleagues examined *Dunaliella salina* and *D. euchlora* from Lake Grasmere, Marlborough, New Zealand. At the lake, commercial salt is produced in ponds by solar means, and the two brine algae are responsible for periodic pink blooms. The data obtained from field studies and laboratory cultures will have value if ever a beta-carotene recovery programme is attempted here.[62]

The filamentous blue-green alga *Calothrix thermalis* plays a strange role in helping to build silica terraces such as the 900 year old White Terraces in the Waiotapu thermal reserve near Rotorua. There seems to be a race between the deposition of silica from the hot water flows, and growth of the alga. Finally, dead strands of *Calothrix* are embedded throughout the silica deposit, with only the emerging tips of the alga still growing at the surface.[9]

The spectacular colours of the Emerald, Rainbow, and Golden Fleece Terraces, and the Artist's Palette area at Orakei Korako geyserland on the Waikato River are due to mats of other blue-green algae, growing in hot water flows with temperatures up to 60°C. Warbrick Terrace in Waimangu thermal valley and the Artist's Palette at Waiotapu have similar colours. Filaments of other blue-greens like *Mastigocladus laminosus* provide the deep green to yellow-green colour, depending on the chemistry and temperature of the habitat, and *Oscillatoria terebriformis* sometimes contributes a red colour, while the flexibacterium *Chloroflexus* at the base of the mat may be orange. In future, it may well be that *Calothrix*, other thermal algae, and silica water will be used to build coloured terraces as landscape features and tourist attractions.[9]

Karerarera (*Azolla filiculoides*) and the introduced water fern (*A. pinnata*) form colourful surface mats on ponds and slow-running streams. The filamentous blue-green alga, *Anabaina*, and some bacteria live in the leaf cavities of *Azolla* species and fix atmospheric nitrogen. Research at the International Rice Institute, Manila, Philippines, includes collecting and crossing different species of *Azolla*, infection of *Azolla* with *Anabaina* and the use of *Azolla* biomass to fertilize crops such as wheat, maize and sunflower. In China, where most late rice paddy is nitrogen deficient, species of *Anabaina* and other blue-green algae such as *Aulosira* and *Nostoc* have been used successfully as biofertilizers.[63]

New Zealand farmers rely on the nitrogen-fixing properties of Rhizobium bacteria in legume root nodules, particularly of the clovers, to obtain nitrogen 'fertilizer' at minimal cost. Blue-green algae might be used however, on land where clovers cannot persist; for example, on South Island high country where seed-pellets were distributed by air to revegetate eroded areas, and the new cover did not survive. Blue-green algae are remarkable for their ability to survive in extreme environments, and Li and Wang (1983) concluded from their studies that all blue-greens can fix atmospheric nitrogen.[64] Other scientists consider that only some blue-greens can fix nitrogen.

Future directions of microalgal research and development could be continued, but enough examples have been given to indicate some of the possibilities of this field. While there are several scientists in New Zealand interested in microalgal taxonomy, there are few salaried posts, and although there are several people trained and experienced in microalgal technology, again there are few openings. Another handicap to progress is the lack of a suitable journal for the publication of work in the related fields of microalgal taxonomy and technology; the few articles by New Zealand authors that do appear come out overseas, or turn up in odd corners of local journals devoted primarily to other disciplines.

Research grants, such as that made by the Commonwealth Government of Australia to develop the *Dunaliella* beta-carotene project, are unknown, and tax incentives to encourage participation are unavailable. In these circumstances, local progress in microalgal research and technology is unlikely to be rapid.

1. Taylor, F.J. (1970) 12(14) *Trans. Roy. Soc. N.Z. Biol. Sci.* 153.
2. Taylor, F.J. (1974a) 4 *J. Roy. Soc. N.Z.* 193.
3. Taylor, F.J. (1974b) 4 *J. Roy. Soc. N.Z.* 395.
4. Wood, R.D., Mason, R. (1977) 15 *N.Z.J. Bot.* 87.
5. Cassie, V. *Revised Checklist of the Freshwater Algae of New Zealand (excluding diatoms and charophytes)* Wellington (NWASCA) 1984, 2 vols.
   Water & Soil Tech. Pub. 25, 1-116p. 1xiv.
   Water & Soil Tech. Pub. 26, 117-250p. 1xiv.
6. Cassie, V. *Checklist of the Freshwater Diatoms of New Zealand* (1984) 4 *Bibliotheca Diatomologica* 1-129.
7. Croasdale, H., Flint, E.A. *Flora of New Zealand Desmids* Vol. 1, Wellington (Government Printer) 1986 133p. 271pl.
8. Croasdale, H., Flint, E.A. *Flora of New Zealand Desmids* Vol. 2, Christchurch (Botany Division) 1988 141p. 61pl.
9. Cassie, V., Cooper, R. *Algae of New Zealand Thermal Areas* Stuttgart (J. Cramer) 1989. Bibliotheca Phycologica Band 78. 159p.
10. Cassie, V. *A Taxonomic Guide to Thermally Associated Algae (excluding Diatoms) in New Zealand* Stuttgart (J. Cramer) 1989. Bibliotheca Phycologica Band 78. p. 161-255.
11. Cassie, V. *A Contribution to the Study of New Zealand Diatoms* Stuttgart (J. Cramer) 1989, Bibliotheca Diatomologica Band 17, 266p. Including 'A pollution tolerance survey of benthic diatoms in New Zealand waterways' p. 1; 'A taxonomic guide to diatoms of thermal habitats in New Zealand, Part 1 — Centrales' p. 51; 'Part 2 — Pennales' p. 94.
12. Bradford, J.M. et al (1986) 20 *N.Z. J. Marine & Freshwater Res.* 253.
13. Chang, F.H., Bradford, J.M. (1985) 19 *N.Z. J. Marine & Freshwater Res.* 193.
14. Gillespie, P.A. (ed.) *Nutrient Processing and Biomass Production in New Zealand Estuaries.* Summaries of contributions to a seminar held at Cawthron Institute, Nelson, 10-11 November 1982, Wellington (NWASCO) 1983. Water & Soil Misc. Pub. 60. 83p.
15. Booth, W.E. (1986) 20 *N.Z.J. Marine & Freshwater Res.* 615.
16. Booth W.E. (1987) 30 *Botanica Marina* 129.
17. Fogg, G.E. (1986) B228 *Proc. R. Soc. Lond.* 1.
18. Jolly, V.H., Brown, J.M.A. (eds) *New Zealand Lakes* (Auckland University Press/Oxford University Press) 1975 388p.
19. Cassie, V. 'The Fossil and Living Freshwater Diatom Flora of New Zealand' In Ross, R. (ed.) *Proceedings of the Sixth Symposium on Recent and Fossil Diatoms*, Budapest 1980 Königstein (Költz) 1981, p. 321.

20. Viner, A.B. (ed.) *Inland Waters of New Zealand* Wellington (Science Information Publishing House) 1987. DSIR Bull 241, 494p.
21. Burns, C.W. 'A note on quantitative phytoplankton studies in some South Island lakes' in *New Zealand Lakes* p. 209.
22. Cassie, V. 'Phytoplankton of Lakes Rotorua and Rotoiti (North Island)', in *New Zealand Lakes* p. 193.
23. Cassie, V. (1978) 12 *N.Z.J. Marine & Freshwater Res.* 153.
24. Flint, E.A. 'Phytoplankton in some New Zealand lakes' in *New Zealand Lakes* p. 163.
25. Round, F.E. (1989) 4(2) *Diatom Research* 70 — editorial.
26. Wilkinson, V. 1981. M.Sc. Thesis, University of Auckland.
27. Cassie, V. (1983a) 11(2) *Arch. Nat. Hist.* 223.
28. Kear, D. (1966) 63 *N.Z. DSIR Information Series* 63.
29. Grange, L.I. (1930) 12 *N.Z.J. Sci. Tech.* 94.
30. Stein, J.R., Borden, C.A. (1984) 23(4) *Phycologia* 485.
31. Cassie, V. (1981) 15 *N.Z.J. Marine & Freshwater Res.* 181.
32. Chang, F.H. (1988) 22 *N.Z.J. Marine & Freshwater Res.* 345.
33. Chang, F.H. (1983) 1 7 *N.Z.J. Marine & Freshwater Res.* 165.
34. Taylor, F.J., Taylor, N.J., Walsby, J.R. (1985) 70(6) *Int. Revue Ges. Hydrobiol.* 773.
35. Chang, F.H., Ryan, K.G. (1985) 24(2) *Phycologia* 191.
36. Taylor, F.J.R. 'Toxic Dinoflagellates' in Raegelis, E.P. (ed.) *Seafood Toxins* 1984; Am. Chem. Soc. Symp. Ser. 262, 460p., p. 77.
37. Hayden, B.J. (1984) 27 *N.Z. Mar. Sci. Soc. Newsletter* 15. (1986) 27 *Catch* 5.
38. (1987) 30 *N.Z. Marine Sciences Society Newsletter* 81.
39. (1988) 31 *N.Z. Marine Sciences Society Newsletter* 44.
40. Boustead, N., Chang, H., Pridmore, R., Todd, P. (1989) *Freshwater Catch* 3; Chang, F.H., Anderson, C., Boustead, N.C. (1990) 24 *N.Z.J. Marine & Freshwater Res.* 461.
41. Okaichi, T., Anderson, D.M., Nemoto, T. (eds) *Red Tides Biology, Environmental Science and Toxicology* New York (Elsevier) 1989 489p.
42. Flint, E.A. (1966) 14(11) *N.Z.J. Vet.* 181.
43. Connor, H.E. *The Poisonous Plants in New Zealand* Wellington (Government Printer) 1977, 2nd ed., 247p.
44. Cassie, V. 'Algae in relation to water quality', in Mulcock, P.A. (ed.) *A review of some Biological Methods for the Assessment of Water Quality with special reference to New Zealand* Wellington (NWASCO) 1979. Water & Soil Tech. Pub 18, 46p., p. 21.
45. Joshi, K.R. Gavin, J.B., Wheeler, E.E. (1975) 56(1) *Mycopathologia* 9.
46. Biggs, B.J.F. (1985) 21(1) *Soil & Water* 27.
47. Coffey, B.T., Miller, S. 1986, pers. commun.
48. *New Zealand Herald* 19 April 1989, Section 1, p. 15.
49. Kokich, D. (1986) 22 *N.Z. Limnological Society Newsletter* 20.
50. Haughey, A. (1968) 2 *N.Z.J. Marine & Freshwater Res.* 721.
51. Haughey, A. (1969) 3 *N.Z.J. Marine & Freshwater Res.* 245.
52. Haughey, A. (1970) 5(1) *Brit. Phycol. J.* 97, 969.
53. Haughey, A. 'Effect of Wastewater Treatment on Inland Waters', 43rd ANZAAS Congress, Auckland, 1979, Abstract 341, 412.
54. Cassie, V. (1983b) 29 *Tane* 119.
55. Burns, C.W. (1979) 18 *N.Z. Water & Soil Tech. Pub.* 5.

56. Bayly, I.A.E. (1989) 23 *N.Z.J. Marine & Freshwater Res.* 357.
57. Tortell, P. 'Non-technical constraints on the development of Aquaculture in New Zealand', Symp. on Coastal Aquaculture, Cochin 1980, Marine Biological Association of India.
58. Roughan, P.G. (1989) 47 *J. Sci. Food Agric.* 85.
59. Borowitzka, L.J. 'Algal Biotechnology in Australia' (1984) 4(2) *Newsletter of the Australasian Society for Phycology and Aquatic Botany* 2.
60. Borowitzka, L.J., Borowitzka, M.A., Moulton, T.P. (1984) 116/117 *Hydrobiologia* 115.
61. Borowitzka, M.A., Borowitzka, L.J. (eds) *Micro-algal Biotechnology* Cambridge University Press, 1988 477p.
62. Chang, F.H., Wear, R.G., Reynolds, J. (1986) 20 *N.Z. J. Marine & Freshwater Res.* 467.
63. (1986) 2(1) *Azolla Newsletter* 1, and subsequent issues.
64. Li (Ley), S.H., Wang, Q. In Tseng, C.K. (ed.) *Proceedings of the Joint China-U.S. Phycology Symposium.* Beijing (Science Press) 1983 336p., p. 479.
65. *New Zealand Herald*, 8 November 1990, Section 1, p. 10. 'Shearwaters starved to death says report'.
66. Taylor, F.J. (1981) 66(3) *Int. Rev. Ges. Hydrobiol.* 377.
67. Taylor, N.J., Taylor, F.J. (1985) 70(2) *Int. Rev. Ges. Hydrobiol.* 173.
68. Mackenzie, A.L., Gillespie, P.A. (1986) 20 *N.Z.J. Marine & Freshwater Res.* 365.
69. Mackenzie, A.L., Kaspar, H.F., Gillespie, P.A. (1986) 20 *N.Z.J. Marine & Freshwater Res.* 397.
70. Thomas, W.H., Scripps Institution of Oceanography, pers. communication, 12 November 1990, (1979) 37 *J. Marine Res.* 327.
71. (1990) 26 *N.Z. Limnological Society Newsletter* 25.
72. Hall, J.A., Vincent, W.F. (1990) 106 *Marine Biology* 465.
73. Desikachary, T.V., Sreelatha, T.V. *Oamaru Diatoms* Berlin Stuttgart (J. Cramer) 1989. Bibliotecha Diatomologica Band 19, 330pp.

15

Bryophytes and Thallophytes
(Excluding Algae)

The role of lower plants in colonizing volcanic rocks and in the formation of soil was described by Leonard Cockayne and other early ecologists as of major importance, but Emanuel Rudolph and Robert Cooper (1953) suggested that dust from a volcanic eruption such as that which occurred at Ngauruhoe in 1949 is far more copious and readily available to pioneer plants than any soil produced by mosses, liverworts, and lichens. However, they do aid plant colonization of bare ground by stabilizing steep soil slopes and soft rock faces and providing a cover in which seeds of higher plants can germinate and grow.[1,2]

15.1 The Bryophytes

Mosses, liverworts, and hornworts belong to a division of the plant kingdom called the Bryophyta, and are the most primitive of greenland plants. The largest are less than 60 cm tall or long, and the smallest are microscopic.

MOSSES
The last systematic treatment of New Zealand mosses was the work of the late G.O.K. Sainsbury (1955)[3]; it is being brought up to date by Allan Fyfe, Botany Institute, Land Resources, DSIR, Lincoln. K.W. Allison and John Child wrote a popular account of New Zealand mosses in 1971[4] with numerous line drawings and photographs. It too is out of date, and out of print, and Jessica Beever, an associate of Botany Institute based at Mount Albert Research Station, has rewritten Allison and Child's book. It will be published shortly. There were 525 species listed as native to New Zealand, and research under way is likely to raise this number.

Soft mosses were used by Māori women in olden times for diapers and to make a soft bed for a baby. Mosses and liverworts used by the Māori for perfumery are mentioned in the section on fragrances. Colenso (1869) referred to them as 'kopura', but this word has other meanings in Williams's *Dictionary of the Maori Language* (1971).[5,6]

SPHAGNUM

Five or six species of the moss genus *Sphagnum* are found in New Zealand, from northern gumfields to southern bogs. *Sphagnum* moss has been used overseas as an absorbent wound-dressing, and there was interest in New Zealand supplies at the outbreak of the Second World War.

There is a frequently told story that a stall holder at a trade exhibition of New Zealand goods in Japan several years ago covered part of the display area with *Sphagnum* as nothing else was available to fill the gap. He received a $400,000 order for twelve container loads of the moss, and a new export industry began.[52] It was soon found that vehicles used to collect *Sphagnum* damage the bogs, and larger operators now employ helicopters to get the wet moss to the drying sheds. One and a half tonnes of wet moss yield about 120 kg of dry moss.

The Westland moss industry exports about 700 tonnes of dried *Sphagnum* annually, earns around ten million New Zealand dollars each year, and employs 400 full-time workers as well as many part-time pickers.[54] Some operators own wetlands where *Sphagnum* grows; others tender for a twelve month licence to pick an area of Department of Conservation or Landcorp land and pay a royalty for each wet tonne of moss. Japanese florists prefer strands of *Sphagnum cristatum* about 15 cm long with good green colour. It must be free of other plants and of broken and discoloured pieces.

A *Sphagnum* moss plant has many large empty cells, and can take in and hold large amounts of water. Also, it has antibiotic properties and tends to keep stored water acidic and sterile. These characteristics make the moss an excellent medium on which to grow young orchid plants, a valuable component of potting mixes for propagating plants other than orchids, and a light, water-holding material for florist's work. The bulk of the New Zealand supply is sold to Japan, with smaller quantities going to South Korea, Taiwan, Hawaii, the US, and Australia.[7,8]

Previously it was doubted whether *Sphagnum* harvesting would be sustainable in Westland. It was thought that supplies would be 'mined' out in ten years, although Ministry of Agriculture staff had tried to speed up regrowth of the moss with a variety of fertilizers.[8] Also gorse was proving a major pest in taking over harvested areas. Since 1987, the DSIR, MAF, and DOC have had budget cuts, staff reductions, and departmental reorganization. Research activity has diminished, and the sustainability of Westland's *Sphagnum* industry remains a mystery. However, Bryan Thomas, a moss collector of Karamea, is one who has tried to develop methods that will ensure sustainable harvesting; he has written a short report on his work. He has experimented with shading and manuring to hasten regrowth after harvesting, tried seeding bare areas with *Sphagnum* dust from the driers, and invented a lighter vehicle for working in the bogs. On its formation in 1987, the Department of Conservation placed a ban on moss collecting from DOC land until it worked out a policy for sustainable harvesting. Since 1987, it has granted licences for more than 3700 hectares of bog. DOC policy is to restrict access to areas of high conservation value and to grant licences only for areas modified by mining, logging, and grazing. Licences are granted for terms longer than one year to encourage collectors to care for the bog and not mine it out as quickly as possible.[55]

When *Sphagnum* harvesting spread from Westland to Otago and Southland recently, concern was expressed at the potential for ecological damage. *Sphagnum* bog occupies sites which have a constant water level; any fluctuation in the water table enables other plants such as rushes, sedges, and cushion plants to enter the bog. On the West Coast, there is usually a surplus of rainwater. Consequently, most soils are too wet and *Sphagnum* flourishes. In Otago and Southland, where rainfall is less, *Sphagnum* bogs play an important part in holding rainwater and releasing it over a period of time. However, these functions have not been measured accurately, and the significance of bogs, especially in mountainous districts, requires further study.

The Protected Natural Areas survey mentioned earlier showed the need for more wetland and tussock reserves in the South Island, and the start of *Sphagnum* moss harvesting in Otago and Southland makes research and conservation increasingly urgent.[54]

PEAT

In the past, *Sphagnum* was left to die and, in time, it formed peat. When the Soil Conservation and Rivers Control Council requested a survey of peat resources in 1976, Professor J.D. McCraw, University of Waikato, Hamilton, co-ordinated fieldwork during the next two years. The survey[9] studied and classified sixty-seven peat areas (excluding Wellington, Stewart Island, and the Chatham Islands, which were the subject of separate reports). Peat was estimated to cover 1400 km^2 and to have a volume of 300 billion m^3.

The peatland types recognized in the survey included moss peats, herb peats, wood peats, and peats with some algae in them. Moss peats consist of species of *Sphagnum* and *Dicranoloma*. Herb peats consist of grass (*'Danthonia'*, *Cortaderia* and *Chionochloa*), sedge (*Carex*), 'reed' (*Typha* and *Phormium*), rush (*Schoenus* and *Baumea*), restiad (*Empodisma* and *Sporodanthus*), fern (*Gleichenia*) and cushion bog (*Oreobolus*). Wood peats include kahikatea (*Dacrycarpus dacrydioides*), bog pine (*Halocarpus bidwillii*), silver pine (*Lagarostrobos colensoi*) and mānuka (*Leptospermum scoparium*). All peat areas showed evidence of interference such as drainage developments, burning, and grazing by stock. Blueberry farming is the latest purpose for which peatland is sought.[9]

Sphagnum moss peat was used for fuel at Waipahi in Southland,[10] and is the preferred type for horticulture, but it is now in short supply. The main source of garden peat for Auckland and Hamilton is at Moanatuatua, where *Empodisma minus* is mined. The common name for it is 'wire-rush', and it is probably better known under its former names *Hypolaena lateriflora* and *Calorophus minor*. It is in the Restionaceae, a southern hemisphere family akin to the rushes.

In Chapter 8 in the section on kauri gum, the mining of Kaimaumau swamp for marketable waxes and resins was mentioned. Another peat mining project, for the production of crude oil, was investigated by the Liquid Fuels Trust Board and Fletcher Challenge Ltd, on the main island of the Chatham Group. If and when the strip mining goes ahead, an investment of $1 billion will be necessary, and the crude oil obtained is estimated to meet 45% of the country's demand. Needless to say, the scheme will be bitterly opposed by conservationists concerned with the fauna and flora of the main island, and is unlikely to be considered further until oil prices rise.

In future, peat types may be in demand for filters to remove industrial wastes from sewage schemes; they already have environmental uses overseas as absorbents for oil spills.

Peat domes form when peat accumulates faster at the centre of a bog than at the edges. Presumably, decay is faster round the edges than it is in the centre. The time scale is indicated by some radiocarbon dates from the South Island, which show that peat formation began 8000 to 10,000 years before the present and after the end of the last major glacial advance. The extent to which peat has been destroyed by draining, clearing, grassing, and fire is seen in the Waikato, where wetland including peat domes once covered most of the district. Now only 15% remains. The 9665 ha Kopouatai peat dome in the Hauraki Plains near Paeroa is the largest and most intact remnant left. The peat is of the restiad type, is 14m deep in places, and developed over the last 12,000 years.[53] Close to Kopouatai peat dome are the wetlands of Whangamarino and the Firth of Thames. The three areas are administered by the Department of Conservation Waikato conservancy, and are the home of thousands of wildfowl and the last refuge for a number of rare plants and animals.

LIVERWORTS
There are 500 species of liverworts and fifteen species of hornworts listed as native to New Zealand. E.A. Hodgson, Kiwi Valley, Hawkes Bay, provided the last systematic treatment in twenty papers published in the *Transactions of the*

Royal Society of New Zealand between 1941 and 1970.[10] K.W. Allison and John Child wrote *The Liverworts of New Zealand* in 1975 as a well-illustrated, popular guide to about two-thirds of the species.[11] Liverworts are sometimes more common than mosses at higher, wetter sites, but have little obvious use. Some species are known to fix atmospheric nitrogen, and one genus, *Schistochila* has been found to cause allergenic contact dermatitis similar to that caused by some higher plants.[12]

15.2 Lichens

'Lichenes' was established as a division of plant classification long before it was realised that lichens are symbiotic associations between fungi and algae in which the fungi have the ability to obtain food from algae living within their tissues. Lichens should be classified according to the characteristics of the fungi but, as many lichens are of doubtful affinity, the division Lichenes is maintained for convenience.

Both *Sticta* and *Pseudocyphellaria* contain species with blue-green algae in their tissues that are capable of fixing atmospheric nitrogen, and nitrogen fixation in lichens is a significant component of the nitrogen balance in some New Zealand forests.[13] Polyporic acid from the lichen *Pseudocyphellaria coronata* was found to prolong the life of leukaemic mice.[14]

In 1985, David Galloway, Botany Division, DSIR, keyed and described the lichens of New Zealand for a volume of the *Flora of New Zealand*. He classified 966 taxa in 210 genera, and considered that they comprised 60% of the lichens to be found here.[15] Since European settlement began in New Zealand, the lichen flora has been reduced drastically in many areas with the clearing and burning of forest and scrub. However, many new habitats have been provided for lichens in exotic forests, and on stone walls, fences, buildings, lamp posts, fruit trees and tarsealed roads.[14]

Lichens have had a place in dyeing for many years, and there are a number of books listing New Zealand species and the mordants which are necessary to obtain certain colours; for example, *Natural Wool Dyes and Recipes* by A. Milner, 1971.[17] David Galloway deplored the use of lichens for dyeing and strongly recommended that they should be conserved. The presence or absence of lichens has been used overseas to indicate environmental pollution. Although they are not affected by acid rain, they are unable to withstand high levels of sulphur dioxide. They have been used to show high fluoride levels near a smelter and to monitor radioactivity. In 1990, two measurement techniques were employed on lichen samples to detect atmospheric pollution. Trace element analysis and electrolyte leakage from cells were used to measure ten elements, and it was concluded that the lichens serve as in situ biomonitors of lead, zinc, boron, and other elements.[18]

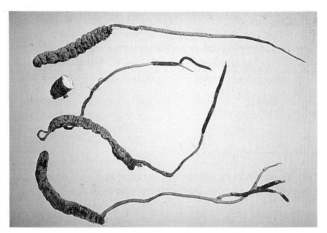

vegetable caterpillar (*Cordyceps robertsii*)
Photo: E. McKenzie

15.3 Fungi

William Hooker of Kew named the first fungus from New Zealand in 1836. It was the remarkable vegetable caterpillar, now known as *Cordyceps robertsii* (Plate 14).[19] Other native fungi have been named since, but Eric McKenzie, Plant Protection, DSIR, Mount Albert, remarked in 1983 that the New Zealand fungal flora was 'still largely undescribed'. Many microscopic groups have not been examined yet and it is possible that the native fungal flora will eventually equal the 12,000 species recorded in Britain.[20]

The Māori recognized about a dozen fungi for food, and these are mentioned in Chapter 8 on Food and Beverages. Dieffenbach (1843) and Colenso (1869) recorded that the Māori used a large species of *Boletus* as tinder and Best (1942) wrote that it was used by travellers in firesticks. Colenso called the fungus 'putawa' and Best referred to it as 'puka tawaii'. The property of slow burning may meet some future need.[5,21]

Colenso sent many dried specimens of fungi to the Reverend M.J. Berkeley in England, and they formed the basis of the section on fungi in J.D. Hooker's *Handbook of the New Zealand Flora*, Volume 2, 1867.[22] The leading New Zealand fungal taxonomist was G.H. Cunningham, Plant Diseases Division, DSIR, Mount Albert, who studied the rust fungi (1931),[50] the gasteromycetes (1944),[23] thelephore (crust) fungi (1963)[24] and polypore (bracket) fungi (1965).[25] Greta Stevenson, Victoria University of Wellington, studied native gilled fungi,[26] R.H. Petersen, University of Tennessee revised the clavarioid fungi (1988),[27] and Egon Horak of Switzerland has completed the first part of an agaric flora of New Zealand (1990).[28]

In the 1977 Banks Lecture to the Royal New Zealand Institute of Horticulture, G.T.S. Baylis, Botany Department, University of Otago, reviewed the beneficial role of mycorrhizas in the native flora. From 1950 onwards, Baylis studied the fungal invasion of root cortices and, with the aid of several research students, devised experiments to show how mycorrhizas aid roots to obtain the phosphorus needed for growth. His studies suggested that heavy top dressing may hinder mycorrhizal plants and thus be wasteful. His work also raised the suspicion that plant breeders have sometimes selected cultivars that do not exploit mycorrhizal symbiosis to the best advantage. One comment by Baylis is repeated here as a hint to home gardeners in northern New Zealand: if your lemon tree is languishing, a spadeful of mycorrhizal soil from a thriving citrus, or from some native bush, may do wonders for it![29]

In September 1988, the British Mycological Society held a symposium 'on the biotechnology of fungi for improving plant growth'. Fungi are being used as agents for the control of weeds, parasitic insects, nematodes, and other fungi. They are also being used to improve plant growth and, here, mutually symbiotic mycorrhizal fungi have an important contribution to make.

Plant Diseases Division (now Plant Protection) DSIR have employed a number of resident plant pathologists, mycologists, bacteriologists and virologists since a Plant Research Station was established at Palmerston North in 1928 as a joint enterprise of the Departments of Agriculture and Scientific and Industrial Research. J.D. Atkinson and colleagues wrote *Plant Protection in New Zealand* (1956) covering about 160 fungal diseases, fifty viral diseases and twenty-five bacterial diseases of plants.[30] By 1969, Joan Dingley had raised the number of plant pathogenic fungi to 700.[31] In 1989 S.R. Pennycook and his colleagues listed more than 6300 New Zealand plant disease records, involving 1102 fungal, seven algal, seventy bacterial and 169 viral pathogens and 1856 host species, mainly garden and crop plants.[32]

Forest tree diseases were recorded separately by Forest Research Institute as forty in 1937, 59 in 1959, and nearly 100 in 1966 (93 fungal, three bacterial, one algal pathogen(s) and one mycorrhizal deficiency disease).[33]

Beneficial activities of soil algae (Flint, E.A.), soil fungi (Thornton, R.H.), yeasts (Di Menna, E.), actinomycetes (Vernon, T.R.) and bacteria (Stout, J.D.) were reviewed in the *Soils of New Zealand*, a three-volume work published by the New Zealand Soil Bureau in 1968. The articles are now out of date; however they remain a clear introduction to the work of these soil organisms.[34-39]

An actinorhizal association with *Frankia* has been recorded for 190 higher plants, including tutu (*Coriaria*) and wild Irishman (*Discaria toumatou*). *Frankia*, the endophyte, has the characteristics of both fungi and bacteria. It fixes nitrogen, an important aid in the invasion of poor soils. Actinorhyzal plants are thought to have a role to play in future conservation work.[40]

tutu (*Coriaria arborea*) **Coriariaceae**
Photo: K.G. Cambie

15.4 Bacteria

Bacteria are usually classified as plants rather than animals, and are regarded with some doubt as relatives of the microscopic algae and fungi. Life on earth would be impossible were it not for the innumerable activities of bacteria. For example, they make dead animal material available as plant food and fix atmospheric nitrogen (of vital importance to plant growth). Denitrifying bacteria in wetland soils have the ability to reduce nitrates to gaseous nitrogen.[41] The Taupo Laboratory of the Division of Water Science, DSIR, has studied the extent to which stream-bank plants and their associated micro-flora are able to remove or transform nutrients originating in fertilized pastures and from sewage effluent outfalls upstream. One result of researches on the removal of nutrients was the decision of the Waikato Valley Authority, as part of the Lake Taupo Catchment Control Commission, to provide buffer strips along streams entering Lake Taupo to reduce the amount of nutrients entering the lake.

Many denitrifying bacteria produce methane under anaerobic conditions. Methane gas is produced and used for energy in a number of sewage treatment plants; for example, at the Manukau sewage treatment plant, Mangere. Methane-producing bacteria are being studied at Cawthron Institute, Nelson, as their energy metabolism is little known.

In *Inland Waters of New Zealand*, 1987, edited by A.B. Viner, W.F. Vincent and D.J. Forsyth of the Taupo Laboratory described some of the archaebacteria and eubacteria found in New Zealand thermal areas. These bacteria live in hot and sometimes acid waters.[42] Their properties are of interest to the Thermophile and Microbial Biochemistry and Biotechnology Unit, established at the University of Waikato, Hamilton, in 1984. Some of the research on thermostable enzymes is contracted to staff at the University of Auckland.[51]

Enzymes with greater chemical and thermal stability can be used in industrial processes with major savings in time and cost. Such enzymes can be achieved through genetic and chemical engineering. It is necessary to establish a record of reliability for such tailor-made enzymes before their novel technology is accepted. Shell New Zealand, which has been involved with DFC and Stratacorp in enzyme studies for about four years, announced in May 1990 that it had taken control from its former partners. One product had been patented, export orders for it had been found and a research contract had been received from an American firm.[43]

An important culture collection was begun by D.W. Dye at Mount Albert Research Station in 1952 as a set of pathogenic bacteria and rhizobia. Fungal cultures were added from 1961 and the collection became international in scope with the addition of authenticated strains from world wide sources. Generally, bacteria cause disease in man much more frequently than fungi, which are of major importance in plant and animal disease.

The 1988 catalogue of the culture collection listed 4400 strains of bacteria and 2400 strains of fungi. Most of the collection is preserved by freeze-drying, and the specimens substantiate New Zealand's disease records, serve as the basis for quarantine decisions, and provide authentic material for a variety of research activities.

15.5 Viruses

R.E.F. Matthews (1981) noted that plant virus diseases found in New Zealand have come mainly from Europe and North America, arriving in corms, tubers, runners, root stocks, seed, and so on. Consequently, they have been recorded in introduced plants, the first in the potato about 1929. The numbers found since then 'reflect the amount of effort put into finding them'. No serious search has been made for viruses in native plants. The vectors for these viruses are introduced species of aphids, and New Zealand is lacking in endemic aphids.[44]

The persistence of human enteroviruses in wastewater and in the environment around sewage outfalls has been noted in the discussion of micro-algae, sewage, and water quality. The studies of Margaret Loutit and co-workers at Dunedin followed the development overseas of techniques for monitoring low viral concentrations. Data linking transmission of viral disease with a waterborne route is limited so far to certain overseas outbreaks of poliomyelitis and infectious hepatitis.

In four papers[46-49] Margaret Loutit and colleagues showed that large amounts of bacteria of faecal origin and human enteroviruses remain in sewage effluent, even after treatment. These disease organisms persist in water and in underlying sediments and appear to accumulate in green-lipped mussels near a sewage outfall. Depuration for eight days (washing in fresh water) did not remove the viruses from the mussels. The results are obviously bad news for swimmers near sewage outfalls, people who use river water downstream from an outfall, and the rapidly growing shellfish industry.

1. Cockayne, L. *New Zealand Plants and their Story* Wellington (Government Printer) 1927 3rd ed. 271p.
2. Rudolph, E.D., Cooper, R.C. (1953) 34 *Ecology* 805.
3. Sainsbury, G.O.K. *A Handbook of the New Zealand Mosses* Wellington (Government Printer) 1955, Roy. Soc. N.Z. Bull. 5, 490p.
4. Allison, K.W., Child, J. *The Mosses of New Zealand* Dunedin (University of Otago Press) 1971 155p.
5. Colenso, W. (1869a, 2nd ed. 1875) 1 *Trans. N.Z. Inst.* 233.
6. Williams, H.W. *A Dictionary of the Maori Language* Wellington (Government Printer) 1971 7th ed. 507p.
7. *New Zealand Herald* 1 February 1989, p. 83, 'Sphagnum moss. Last laugh goes to swamp buyer'.
8. *New Zealand Farmer* 25 April 1985, p. 16, 'Sphagnum moss. Resources shrink'.
9. Davoren, A. *A survey of New Zealand Peat Resources.* Hamilton (University of Waikato for NWASCO) 1978, Water & Soil Tech. Pub. 14 157p.
10. Hodgson, E.A. (1941) 71 *Trans. Roy. Soc. N.Z.* 181, and subsequent papers to 1970.
11. Allison, K.W., Child, J. *The Liverworts of New Zealand* Dunedin (University of Otago Press) 1975 300p.
12. Anon. (1988) 11 *N.Z. Bot. Soc. Newsletter* 5.
13. Green T.G.A., Horstmann, J., Bonnett, H., Wilkins, A., Silvester, W.B. (1980) 84 *New Phytologist* 339.
14. Burton, J.F., Cain, B.F. (1959) 184 *Nature* 1326.
15. Galloway, D.J. *Flora of New Zealand: Lichens* Wellington (Government Printer) 1985 662p.
16. Hayward, B.R. (1985) *Bot. Div. Newsletter Suppl.* 2, p.13.
17. Milner, A. *Natural Wool Dyes and Recipes* Dunedin (John McIndoe) 1971 33p.
18. Rope, S.K., Pearson, L.C. (1990) 93(1) *The Bryologist* 50.
19. Hooker, W.J. (1836) *Icones Plantarum* t. 11 *Sphaeria robertsii.*
20. McKenzie, E.H.C., Galloway, D.J. In Brownsey, P.J., Baker, A.N. (eds) *The New Zealand Biota — what do we know after 200 Years* National Museum of New Zealand Misc. Ser. 7/91p., p. 7, 1983.
21. Dieffenbach, E. *Travels in New Zealand* London (J. Murray) 1843, 2 vols. 431 396p, Capper Press reprint 1974.
22. Hooker, J.D. *Handbook of the New Zealand Flora* London (Lovell Reeve) 2 vols; vol. 1 1864 392p; Vol. 2 1867 p. 393-798.

23. Cunningham, G.H. *The Gasteromycetes of Australia and New Zealand* Dunedin (John McIndoe) 1944 236p.
24. Cunningham, G.H. *The Thelephoraceae of Australia and New Zealand* Wellington (Government Printer) 1963, DSIR Bull 145, 359p.
25. Cunningham, G.H. *Polyporaceae of New Zealand* Wellington (Government Printer) 1965, DSIR Bull. 164, 304p.
26. Stevenson, G. (1962) 16 *Kew Bull.* 373, and other papers.
27. Petersen, R.H. *The Clavarioid Fungi of New Zealand* Wellington (S.I.P.C.) 1988 DSIR Bull. 236, 170p.
28. Horak, E. (1990) 28 *N.Z. J. Bot.* 255.
29. Baylis, G.T.S. (1978) 6 *Ann. J. Roy. N.Z. Inst. Hort.* 63.
30. Atkinson, J.D., Chamberlain, E.E., Dingley, J.M., Reid, W.D., Brien, R.M., Cottier, W., Jacks, H., Taylor, G.G. *Plant Protection in New Zealand* Wellington (Government Printer) 1956 699p.
31. Dingley, J.M. *Records of Plant Diseases in New Zealand* Wellington (Government Printer) 1969, DSIR Bull. 192, 298p.
32. Pennycook, S.R. *Plant Diseases Records in New Zealand* Auckland (Plant Diseases Division, DSIR) 1989, 3 vols.
    vol. 1 Host list of plant diseases in New Zealand, 276p.
    vol. 2 Fungal plant diseases, 502p.
    vol. 3 Algal plant diseases; Bacterial plant diseases by Pennycook, S.R, Young, J.M., Fletcher, M.J. Viral plant diseases by Pennycook, S.R., Fry, P.R., 150p.
33. Gilmour, J.W. *The Pathology of Forest Trees in New Zealand. The Fungal, Bacterial, and Algal Pathogens* 1966 FRI, NZFS Tech. Paper 48 82p.
34. New Zealand Soil Bureau *Soils of New Zealand* Wellington 1968 N.Z. Soil Bureau Bull. 26, 3 vols.
35. Flint, E.A. 'Algae on the surface of some New Zealand soils' in N.Z. Soil Bureau Bull 26, 1968, Vol. 2, p. 183.
36. Thornton, R.H. 'Fungi in New Zealand Soils' N.Z. Soil Bureau Bull 26, 1968, Vol. 2, p. 191.
37. Di Menna, M.E. 'Yeasts [in New Zealand Soils]' N.Z. Soil Bureau Bull 26, 1968, Vol. 2, p. 195.
38. Vernon, T.R. 'Actinomycetes [in New Zealand Soils]' N.Z. Soil Bureau Bull 26, 1968, Vol. 2, p. 202.
39. Stout, J.D. 'Bacterial Flora' and 'Microfauna' N.Z. Soil Bureau Bull 26, 1968, Vol. 2, p. 202.
40. Bullock, B.T. 14 *Annual Journal RNZIH* 33 (1986/7).
41. Bartlett, M.S., Brown, L.C., Hanes, N.B., Nickerson, N.H. (1979) 8(4) *J. Environmental Quality* 460.
42. Vincent, W.F., Forsyth, D.J. 'Geothermally influenced waters' in Viner, A.B. (ed.) *Inland Waters of New Zealand.* Wellington (Science Information Publishing Centre) 1987 DSIR Bull 241, 494p.
43. *New Zealand Herald* 18 May 1990 'Enzyme activity'.
44. Matthews, R.E.F. *Plant Virology* New York (Academic Press) 1981, 2nd ed. 897p.
45. Brock, T.D. *Thermophiles — general, molecular, and applied microbiology* New York (Wiley Interscience) 1986 316p.
46. Lewis, G., Loutit, M.W., Austin, F.J. (1986) 20 *N.Z.J. Marine & Freshwater Research* 431.
47. Lewis, G.D., Austin, F.J., Loutit, M.W. (1986) 20 *N.Z.J. Marine & Freshwater Research* 101.

48. Lewis, G.D., Loutit, M.W., Austin, F.J. (1985) 19 *N.Z.J. Marine & Freshwater Research* 187.
49. Loutit, M.W., Lewis, G., (1985) 19 *N.Z.J. Marine & Freshwater Research* 179.
50. Cunningham, G.H. *The Rust Fungi of New Zealand* Dunedin (J. McIndoe) 1931, 261p.
51. Daniel, R.M., Morgan, H.W., Hudson, J.A. (1987) 113 (1548) *New Scientist* 36.
52. *New Zealand Herald* 20 July 1984. 'Last minute idea won the day'.
53. De Lange, P. 1989 M.Sc. Thesis, University of Waikato.
54. Johnson, P. (1988) 'Sphagnum Moss in Otago, A discussion report'. Wellington (Department of Conservation, Science & Research Directorate) S & R Internal Report 34, 18p.
55. Yarwood, V. (1990) 7 *New Zealand Geographic* 55.
56. Thomas, B. (1984) 'Moss farming in Karamea. A report on the development of farming practices and techniques based on observations made over the last four years' Karamea (Thomas) 32p.

Index

Trifolium repens L., white clover 111
Trisetum antarcticum Trinius 88
Trisetum subspicatum — now *T. spicatum* L. Richter 88
Triticum sativum L., wheat 99, 197
Trunk wood 61, 62
Tūpākihi — see *Coriaria*
Tussock 12, 14, 67, 85, 87, 88, 90, 91, 92, 95
Tussock, blue — see *Poa colensoi*
Tussock, broad-leaved snow — see *Chionochloa flavescens*
Tussock, narrow-leaved snow — see *Chionochloa rigida*
Tussock, Nasella — see *Nasella trichotoma*
Tussock, red — see *Chionochloa rubra*
Tussock, tall — see *Chionochloa*
Tūtae kēhua — see *Clathrus cibarius*
Tutu — see *Coriaria*
Typha orientalis C.B. Presl, bullrush, raupō, reed 9, 83, 100, 110, 203

Ulva lactuca L., sea lettuce 108
Uncinia uncinata (L.F.) Kirk, 'Copperware shrub' 49
Undaria 183
Undaria pinnatifida (Harvey) Suringar, wakame 183
Urtica ferox Forst. f., ongaonga, tree nettle 122, 126, 127
Urtica incisa Poiret, dwarf bush nettle 102
Urtica urens L., nettle 122

Vascular plants 1
Vegetable caterpillar — see *Cordyceps robertsii*
Vegetable sheep 1, 48
Verbenaceae **Plate 13**
Veronica — see *Hebe* 56
Veronica 'Blue Gem' — now *Hebe x franciscana* cv. 'Blue Gem'
Veronica decussata — now *Hebe elliptica*
Veronica lobelioides = *V.* 'Blue Gem' Hort. 33
Veronica lyallii — now *Parahebe lyallii*
Veronica 'Purple Queen' Hort. 33
Veronica salicifolia — now *Hebe stricta* and allied species
Veronica speciosa — now *Hebe speciosa*
Viper's bugloss — see *Echium vulgare*
Virus diseases 207, 209, 210
Virus diseases, yellow leaf of N.Z. flax 80
Vitex lucens Kirk, pūriri 8, 10, 27, 61, 70, 71, 134, **Plate 13**

Waewae atua — see *Dactylanthus taylorii*

Waewaekoukou — see *Lycopodium volubile*
Wai kōrari, flax honey 160
Wakame — see *Undaria pinnatifida*
Waoriki — see *Ranunculus amphitrichus*
'Water cress' — possibly *Rorippa* species
Water fern — see *Azolla pinnata*
Wāwā — see *Scirpus lacustris*
Weeds 14, 87, 94, 95, 207
Weinmannia racemosa L.f., kāmahi, tōwai 8, 85, 111, 112, 134, 135, 136
Weinmannia silvicola A. Cunn., tawhero 111, 112, 134, 136
Whākou — see *Ixerba brexioides*
Wharangi — see *Melicope ternata*
Wharariki — see *Phormium cookianum*
Whauwhi — see *Plagianthus regius*
Wheat — see *Triticum sativum*
Whekī — see *Dicksonia squarrosa*
White clover — see *Trifolium repens*
White pine — see *Dacrycarpus dacrydioides*
White teatree — see *Kunzea ericoides*
Wī, grass 87
Willow 16
Willow herb — see *Epilobium*
Wine — see *Aristotelia* and *Coriaria*
Wineberry — see *Aristotelia serrata*
Winter's bark — see *Pseudowintera*
Wire rush — see *Empodisma minus*
Wood 61, 62

yam, 'yamms' — see *Dioscorea alata*
yeast 125, 132, 207
yellow leaf disease of N.Z. flax 80
Yorkshire fog — see *Holcus lanatus*

Zanzibar weed — see *Eucheuma 'cottonii'* var.
Zea mays L., maize 46, 107, 197
Zostera, seagrass 12, 83
Zoysia 94

GENERAL

Acland, J.B.A. 47
Adams, Olga 120
Agar, 14, 177–79, 181
Aiton, William 19
Akaroa plants 38
Allan, H.H. 14, 15, 31, 89, 90
Allison, K.W. 201, 205
Alpine plants 29–30, 48
Ambergris 155–56
Anderson, William 7, 20–21

CHEMICAL NAMES